U.S.NRC

United States Nuclear Regulatory Commission

Protecting People and the Environment

NUREG-1886

I0502884

Joint Canada – United States Guide for Approval of Type B(U) and Fissile Material Transportation Packages

Final Report

Manuscript Completed: March 2009
Date Published: March 2009

Prepared by:
M. Rahimi, N.L. Osgood, and M. Sampson[1]
M. Conroy[2]
S. Faille and K. Glenn[3]

[2]U.S. Department of Transportation
Pipeline and Hazardous Materials Safety Administration
1200 New Jersey Avenue, SE
Washington, DC 20590

[3]Canadian Nuclear Safety Commission
Packaging and Transport Licensing Division
P.O. Box 1046
280 Slater Street
Ottawa, Ontario K1P 5S9

M. Sampson, NRC Project Manager

[1]Office of Nuclear Material Safety and Safeguards

ABSTRACT

This guide was developed by the United States Department of Transportation and Nuclear Regulatory Commission, and the Canadian Nuclear Safety Commission to provide direction on the standard format and content for applications for approval of packages used to transport Type B(U) and fissile materials. The guide is designed to facilitate the United States validation of Competent Authority approvals for export/import purposes and limit redundant technical reviews. A companion guide document has been developed and published in Canada as RD-364, "Joint Canada - United States Guide for Approval of Type B(U) and Fissile Material Transportation Packages."

This document is intended for use by package approval applicants to assist in preparing applications that thoroughly demonstrate the ability of the given package to meet both United States and Canadian regulations. The guide applies specifically to applications for approval of Type B(U) and fissile material (Type A and Type B) transportation packages intended for export/import. The guide describes a method that is acceptable to the staff of the United States Nuclear Regulatory Commission, United States Department of Transportation, and the Canadian Nuclear Safety Commission.

Nothing contained in this guide is to be construed as having the force or effect of regulations. Comments regarding errors or omissions, as well as suggestions for improvement of this NUREG should be sent to the Director, Spent Fuel Storage and Transportation Division, U.S. Nuclear Regulatory Commission, Washington, D.C. 20555-0001.

CONTENTS

PREFACE

The International Atomic Energy Agency's (IAEA) "Regulations for the Safe Transport of Radioactive Material" (Safety Standards Series No. TS-R-1), specifies that once a Type B(U) transportation package design has been approved by one country, it can be used in other countries without additional review. In actuality, however, some Member States, including Canada and the United States, review all applications for package design prior to allowing its use in their country.

In Canada and the United States, competent authorities review all foreign-certified package designs before issuing the competent authority approval (revalidation). This process can be lengthy and is complicated by differences in domestic regulations, interpretation of IAEA regulations, package application format, and acceptance criteria.

The Canadian Nuclear Safety Commission (CNSC), the United States Nuclear Regulatory Commission (NRC), and the United States Department of Transportation (DOT) have cooperated to produce this guide to facilitate the Canadian and United States regulatory approvals of Type B(U) and fissile package design certificates. This guide assists applicants in preparing applications that thoroughly and completely demonstrate the ability of the given package to meet either Canadian or United States regulations, as applicable.

This document describes a method that is acceptable to the CNSC for complying with Canada's "Packaging and Transport of Nuclear Substances (PTNS) Regulations" (SOR/2000-208), which incorporate, in part, IAEA TS-R-1, 1996 Edition (Revised), and is acceptable to DOT and the NRC for complying with the United States regulations in Title 10, Part 71, "Packaging and Transportation of Radioactive Materials," of the *Code of Federal Regulations* (10 CFR Part 71).

This guide is published in Canada as RD-364, "Joint Canada-United States Guide for Approval of Type B(U) and Fissile Material Transportation Packages." In the United States, it is published as NUREG-1886, "Joint Canada-United States Guide for Approval of Type B(U) and Fissile Material Transportation Packages." RD-364 pertains to Canadian regulatory requirements, while NUREG-1886 pertains to United States regulatory requirements.

Nothing contained in this document is to be construed as relieving any applicant from the requirements of any pertinent regulations. It is the applicant's responsibility to identify and comply with any applicable legislation, regulation or standards.

x

ABBREVIATIONS

ALARA	as low as reasonably achievable
ANS	American Nuclear Society
ANSI	American National Standards Institute
ASME	American Society of Mechanical Engineers
Bq	becquerel
C	Celsius
CFR	*Code of Federal Regulations*
Ci	curie
CNSC	Canadian Nuclear Safety Commission
CSI	criticality safety index
DOT	U.S. Department of Transportation
F	Fahrenheit
GWD/MITHM	gigawatt-days per metric ton initial heavy metal
H/X	hydrogen-to-fissile atomic ratio
IAEA	International Atomic Energy Agency
ICAO	International Civil Aviation Organization
ISO	International Standardization Organization
k_{eff}	effective neutron multiplication factor
kg	kilogram
kPa	kilopascals
lbf/in.^{2}	pound-force per square inch
LDM	low dispersible material
MeV	megaelectronvolts
MPa	megapascal
mrem	millirem
MWD/MTU	Megawatt-days per metric ton uranium
NQA-1	ANSI/ASME, "Quality Assurance Program Requirements for Nuclear Power Facilities," 1983 Edition
NRC	U.S. Nuclear Regulatory Commission
PTNS	Canada's "Packaging and Transport of Nuclear Substances Regulations," SOR/2000-208
QA	quality assurance
SI	international system of units
SNF	spent nuclear fuel
Sv	Sievert
TBq	terabecquerel
TS-R-1	IAEA, "Regulations for the Safe Transport of Radioactive Material," 1996 Edition (Revised)

A. INTRODUCTION

This guide provides a standard format and content for applications for approval of Type B(U) and fissile material (Type B and Type A) transportation packages. The objective is to facilitate the process of certification and subsequent validation by Canadian and United States (U.S.) competent authorities.

For international shipments, the package design must be certified by the competent authority of the originating country, either the Canadian Nuclear Safety Commission (CNSC) or the United States Department of Transportation (DOT). For packages to be certified and validated under this joint Canada – United States guide, the package application format and content guidance in this document should be used.

In Canada, package designs for both for domestic and international shipments must be certified by the CNSC. When CNSC-only certification is sought, other application formats may be acceptable providing the requirements of the "Packaging and Transport of Nuclear Substances (PTNS) Regulations", SOR/2000-208 (Ref. 1) are met.

In the U.S., package designs for both domestic and international shipments are certified by the Nuclear Regulatory Commission (NRC). DOT issues a companion certificate (Competent Authority Certificate) for an NRC-certified package design that is used for U.S. import and export shipments. When only NRC certification is sought, the NRC's standard format or another format of the applicant's choice is acceptable providing the requirements of the NRC's regulation at Title 10, Part 71 "Packaging and Transportation of Radioactive Materials" of the *Code of Federal Regulations* (10 CFR Part 71) (Ref. 2) are met.

For packages with unique design features, the DOT exercises its option to have the NRC perform technical reviews for import/export purposes. Similarly, the CNSC also performs technical reviews for these packages.

This guide describes a method that is acceptable to the CNSC for complying with the Canadian PTNS Regulations, which are based on the International Atomic Energy Agency (IAEA), Regulations for the Safe Transport of Radioactive Material," Safety Standards Series No. TS-R-1, 1996 Edition (Revised) (Ref. 3), and is acceptable to the DOT and the NRC for complying with the U.S. requirements at 10 CFR Part 71. This guidance is not intended as an interpretation of these regulations.

This guide is published in Canada as RD-364, "Joint Canada-United States Guide for Approval of Type B(U) and Fissile Material Transportation Packages." In the United States, it is published as NUREG-1886, "Joint Canada-United States Guide for Approval of Type B(U) and Fissile Material Transportation Packages." RD-364 pertains to Canadian regulatory requirements, while NUREG-1886 pertains to United States regulatory requirements.

Adherence to this guide does not preclude competent authorities of both countries from performing a more detailed technical review of any application.

Nothing contained in this guide is to be construed as having the force or effect of regulations, or as indicating that applications supported by safety analyses and prepared in accordance with the recommendations of this guide will necessarily be approved, or as relieving any applicant or certificate holder from the requirements of any pertinent regulations. This guide cites applicable sections of the regulations, and the applicant should review the cited regulatory requirements to ensure full understanding of the requirement.

B. PURPOSE

This guide assists applicants in preparing applications that thoroughly and completely demonstrate the ability of the given package to meet either Canadian or United States regulations, as applicable. It is also intended to assist reviewers in the review and approval of applications. Where differences in the regulatory requirements exist, guidance is provided to assist the applicant in appropriately addressing the specific regulatory requirement.

C. SCOPE

1. General Information

This guide applies specifically to applications for approval of Type B(U) and fissile material (Type B and Type A) transportation packages in accordance with the NRC and CNSC package requirements. This guide does not apply to approval of special form materials, certain air shipments of Type B packages, low dispersible material (LDM), Type C packages, or fissile material in less than Type A packages.

2. Air Shipments

For shipments of Type B packages by air, the limits on the content specified in the International Civil Aviation Organization (ICAO) regulations must be met for international shipments and for shipments in Canada. In order to transport radioactive material in a package in quantities exceeding the limits for Type B packages specified in the ICAO regulations, the use of a Type C package is required or demonstration that the material conforms to the requirements for LDM is needed.

The NRC regulations include specific provisions for the transport of plutonium by air. These regulations are 10 CFR 71.64, "Special Requirements for Plutonium Air Shipments," and 10 CFR 71.88, "Air Transport of Plutonium," which, among other things, specify the packages and shipments of plutonium that are affected, and 10 CFR 71.74, "Accident Conditions for Air Transport of Plutonium," which specifies the test conditions for plutonium packages transported by air.

CNSC regulations include requirements that limit the quantity of radioactive material that can be transported by air in a Type B package (unless the material is certified as LDM). The CNSC requirements are not specific to plutonium but are applicable to all radioactive material greater than 3,000 A_1 or 100,000 A_2 (whichever is lower) for special form material and 3,000 A_2 for non-special (normal) form material (Table I of TS-R-1 defines A_1 and A_2 activity limit values, which are derived in Section IV of the standard). Therefore, applications for approval of air shipments of plutonium, materials above these values, and Type C packages are outside the scope of this document.

3. Low Dispersible Material Applications

LDM requires multilateral approval, and a Type B package carrying LDM also requires multilateral approval under the IAEA TS-R-1 regulations. Approval for LDM material is required from Canada and the United States in accordance with TS-R-1. Application for approval of LDM is outside the scope of this document.

4. Type B(U) Packages

Package designs to contain Type B quantity radioactive material for roadway, railway, and seaway are within the scope of this guide. The packages should demonstrate the ability to meet the requirements for Type B(U) packages for approval in Canada and the United States, as described in this guide.

5. Fissile Material Packages

Package designs to contain fissile material should demonstrate the ability to meet the requirements for Type AF or Type B(U)F packages for approval in Canada and the United States, as described in this guide. U.S. regulations do not recognize industrial package designs for fissile material.

D. APPLICATION PROCESS

1. General Information

The application should consist of the safety analysis report and the approval documents issued by Canada or the United States, if available, and a cover letter stating the applicant's name, address, fax number or email address, and telephone number; purpose of the application; and mode of transport required. Canadian applicants should submit their applications to the Canadian competent authority for initial approval and to the United States competent authority for subsequent validation. United States applicants should submit their applications to the NRC for Type B and fissile material packages for initial approval and to the Canadian competent authority for subsequent validation.

2. Addresses

The application should be submitted to the appropriate regulatory authority.

For Canadian packages and validation of non-Canadian packages:

Canadian Nuclear Safety Commission
Transport Licensing and Strategic Support Division
P.O. Box 1046, Station B
280 Slater Street
Ottawa, Ontario K1P 5S9

Fax: (613) 995-0556
Phone: 1-888-229-2672
Email: Transport@cnsc-ccsn.gc.ca

For United States validation of certificates for packages to be used for import or export shipments between the United States and Canada:

U.S. Department of Transportation
Radioactive Materials Branch
Office of Hazardous Materials Technology
Pipeline and Hazardous Materials Safety Administration
East Building, E21-303, PHH-23
1200 New Jersey Avenue, SE
Washington, DC 20590

Fax: (202) 366-3753
Phone: (202) 366-4545
Email: ramcert@dot.gov

For U.S. approval of Type B and fissile material packages for U.S. domestic shipment:

U.S. Nuclear Regulatory Commission
ATTN: Document Control Desk
Director, Division of Spent Fuel Storage and Transportation
Office of Nuclear Material Safety and Safeguards
Washington, DC 20555-0001

Fax: (301) 492-3300
Phone: (301) 492-3345
Email: EIE@nrc.gov

Unless requested otherwise, all information submitted to the respective competent authorities will be available for public disclosure. Proprietary information, such as specific design details shown on certification drawings, may be withheld from public disclosure.

For the U.S., the applicant's request for withholding must be accompanied by an affidavit and must include information to support the claim that the material is proprietary in accordance with U.S. regulations at 10 CFR 2.390 "Public Inspections, Exemptions, Requests for Withholding" (Ref. 4).

For Canada, CNSC is subject to the *Access to Information Act* (http://laws.justice.gc.ca/en/a-1/218072.html). The Act provides a right of access to records under the control of a government institution. All information is released in response to requests unless one or more of the limited and specific exceptions to the right of access apply. CNSC has an obligation to release all information that can reasonably be severed from that information which qualifies for an exception. (CNSC will not blanket exempt a record if only part of the information qualifies for an exception.) It is the applicant's responsibility to prove that the information in a record meets the requirements set forth in each claimed exception. CNSC decisions to deny access to information may by reviewed independently by the Office of the Information Commissioner of Canada and by the Federal Court of Canada.

3. Schedule for Submittals

In general, the application should be submitted at least 1 year in advance for new designs, 6 months for amendments, and 3 months for renewals and revalidations before approvals are required.

Applicants should give sufficient advance notice to the respective competent authorities before any design confirmatory tests. Applicants are encouraged to meet with the regulatory body before and during the design process to facilitate a clear understanding of the regulatory requirements.

E. APPLICATION FORMAT

1. General Information

The application is the principal document in which an applicant provides the information and bases for the respective competent authority's staff to use in determining whether a given package meets the requirements of the respective country's package standards. Toward that end, this guide identifies the information to be submitted and provides a format for presenting that information when both United States and Canadian approval is requested.

When CNSC-only certification is sought, other application formats may be acceptable providing the requirements of the PTNS Regulations are met. When only NRC certification is sought, the NRC's standard format or another format of the applicant's choice is acceptable as long as the requirements of the NRC's regulation, 10 CFR Part 71, are met.

In addition to this guide, the information provided in the application should be current with respect to the state of technology for transportation of radioactive materials and should account for any recent changes in the competent authorities' regulations and guides; industry codes and standards; developments in transportation safety; and experience in the design, construction, and use of packages for radioactive materials.

Applicants should strive for clear and concise presentation of the information required in the application. Confusing or ambiguous statements and unnecessarily verbose descriptions do not contribute to expeditious technical review. Claims regarding the adequacy of designs or design methods should be supported by technical bases (i.e., an appropriate engineering evaluation or description of actual tests). Terms should be used as defined in the packaging and transportation regulations.

The safety analysis report should follow the numbering system and headings of the format to at least the third level (e.g., 2.1.2, "Design Criteria") as shown in Section F of this guide. When a particular requirement does not apply to a given package, applicants should use the phrase "Not Applicable," rather than omitting the corresponding section. In addition, applicants should offer a reason for not addressing a particular requirement when its applicability is questionable.

Appendices to each section of an application should include detailed information omitted from the main text. The first appendix to a given section of an application should provide a list of documents that are referenced in the text of that section, including page numbers, if appropriate. If an application references a proprietary document, it should also reference the nonproprietary summary description of that document.

Appendices to each section of an application should provide sufficient detail and photographs to support all physical tests of components and packages addressed in the given section. Applicants may also use appendices to provide supplemental information that is not explicitly identified in the standard format.

When an application cites numerical values, the number of significant figures should reflect the accuracy or precision to which the number is known. In addition, the SI units, along with equivalent conventional units, if appropriate, should be provided for the numerical values. When appropriate, the applicant should specify estimated limits of error or uncertainty.

Applicants should not drop or round off significant figures if doing so would inadequately support subsequent conclusions.

Applicants should use abbreviations, symbols, and special terms consistently throughout an application and in a manner that is consistent with generally accepted usage. Each section of an application should define any abbreviations, symbols, or special terms used in the given section that are unique to the proposed packaging or not common in general usage.

Applicants should use drawings, diagrams, sketches, and charts when such means would more accurately or conveniently present the information to be conveyed. Applicants should ensure that drawings, diagrams, sketches, and charts present information in a legible form with relevant symbols defined. However, the details included in the drawings should be at a level that is sufficient for certification purposes. Additional details on certification drawings, similar to those needed for fabrication drawings, will require frequent review and approval by regulatory authorities if any changes are made and may not be necessary to support package certification. In addition, applicants should not reduce drawings, diagrams, sketches, and charts to the extent that readers need visual aids to interpret pertinent information.

Applicants should number pages sequentially within each section and appendix. For example, the fourth page of Section 6 should be numbered 6-4.

2. Electronic Submissions

If an applicant submits all or part of an application electronically, the submission must be made in a manner that enables the agency to receive, read, authenticate, distribute, perform text search, and archive the submission and process and retrieve it one page at a time.

3. Revisions

For hard copies, applicants should update data and text by replacing pages, rather than using "pen and ink" or "cut and paste" changes. For electronic submissions, applicants should submit the updated safety analysis reports in their entirety. In addition, applicants should provide a list of the changes and highlight the updated or revised portion of each page using a change indicator consisting of a bold vertical line drawn in the margin opposite the binding margin. The line should be the same length as the portion actually changed.

All pages submitted to update, revise, or add pages to an application should show the date of the change and the corresponding change or amendment number. A transmittal letter, including a guide page listing the pages to be inserted and removed, should accompany the revised pages. When applicable, supplemental pages may follow the revised page.

All statements on a revised page should be accurate as of the date of each submittal. Applicants should take special care to ensure that they revise the main sections of the application to reflect any design changes reported in supplemental information (e.g., responses to NRC and CNSC staff requests for information).

F. APPLICATION CONTENT GUIDANCE

This chapter provides the recommended format for the safety analysis report. The use of a uniform format will help to ensure the completeness of an application and assist reviewers in locating information.

1. <u>General Information</u>

This section of the application should present an introduction and a general description of the package. The applicant should also specify that the application is intended to meet the regulatory requirements of the CNSC regulations, namely the PTNS regulations, SOR/2000-208 which refer to IAEA TS-R-1, 1996 (Revised), and the NRC's regulations at 10 CFR Part 71. As indicated in this guide, the application must demonstrate that the package meets the more stringent of the two. In the following text, shaded boxes identify the more stringent requirements where there are significant differences. If the approval is only sought for one of the two countries, this guide may not apply.

The purpose of the application should be clearly stated. The application may be for approval of a new design, for modification of an approved design, or for renewal of an existing approval. Applications for a new design should be whole and complete and should contain the information required in Subpart D, "Application for Package Approval," of 10 CFR Part 71 and Section VI of TS-R-1 which is referenced in Subsection 1(4) of the PTNS Regulations, as applicable.

Applications for modification of an approved design should clearly identify the changes being requested. Modifications may include design changes, changes in authorized contents, or changes in conditions of the approval. Design changes should be clearly identified in revised packaging drawings. The application should include an assessment of the changes and a justification that these changes do not affect the ability of the package to meet the regulatory requirements. Applications for modifications may be subject to the provisions of 10 CFR 71.19, "Previously Approved Package," and Paragraphs 816 and 817 of TS-R-1 as referenced in Paragraph 16(1)(a)(viii) of the PTNS Regulations. Amendments are applicable only to package designs that have been approved under the provisions of this guide. For package designs approved before this guide became effective, revised package applications may be submitted for approval under this guide.

Packaging that does not conform to the drawings referenced in the design approval is not authorized for use. Likewise, only contents specified in the approval may be transported. Package operations, acceptance tests, and the maintenance program may also be specified as conditions of the approval.

1.1 Introduction

This section should identify the proposed use of the package, the model number, and, in the case of fissile material packages, the proposed criticality safety index (CSI) and the value of "N," as defined in 10 CFR 71.59, "Standards for Arrays of Fissile Material Packages," and Paragraph 681 of TS-R-1, which is incorporated in Subsection 1(1) of the PTNS Regulations by reference to Paragraph 672 of TS-R-1. This section should clearly specify any restrictions regarding transport mode, stowage, exclusive use, or type of conveyance for shipment of the package.

1.2 Package Description

This section should include a package description as required by 10 CFR 71.33, "Package Description," and Paragraphs 807 and 813 of TS-R-1 as referenced in Paragraph 7(1)(a) of the PTNS Regulations. The package description should be sufficiently detailed to provide an adequate basis for its evaluation.

1.2.1 Packaging

This section should describe the packaging and any ancillary equipment, with the major design features highlighted. Sketches, figures, or other schematic diagrams should be provided as appropriate. Engineering drawings of the packaging should be presented in the appendix. The general packaging description should include the following information:

1.   The overall dimensions (the smallest overall dimension of the package must not be less than 10 centimeters (cm) (4 inches (in.)) as required by 10 CFR 71.43(a) or Paragraph 634 of TS-R-1 which is incorporated in Subsection 1(1) of the PTNS Regulations by reference to Paragraph 650 of TS-R-1;

2.   Maximum (fully loaded) weight and minimum (empty) weight, if appropriate;

3.   Maximum normal operating pressure as defined in 10 CFR 71.4, "Definitions," or Paragraph 228 of TS-R-1;

4.   Structural features, including lifting and tie-down devices, impact limiters or other energy-absorbing features, internal supporting or positioning features, outer shell or outer packaging, and packaging closure devices;

5.   Secondary packaging components, including internal containers, spacers, shoring;

6.   For spent nuclear fuel (SNF), internal components, such as baskets and any inner containers for damaged or consolidated fuel;

7.   Tamper-indicating features as specified in 10 CFR 71.43(b) or Paragraph 635 of TS-R-1 which is incorporated in Subsection 1(1) of the PTNS Regulations by reference to Paragraph 650 of TS-R-1;

8.   Packaging markings (e.g., model number, serial number, gross weight, and assigned identification number);

9.   The codes and standards used for package design, materials specification, fabrication, welding, and inspection;

10.   Heat transfer features;

11.   Containment features, including penetrations such as vents, ports, and sampling ports;

12.   Neutron and gamma shielding features, including personnel barriers; and

13.   Criticality control features, including neutron poisons, moderators, flux traps, and spacers.

The exact containment system boundary should be defined. This may include the containment vessel, welds, drain or fill ports, valves, seals, test ports, pressure relief devices, lids, cover plates, and other closure devices. If multiple seals are used for a single closure, the seal defined as the containment system seal should be clearly identified. A sketch of the containment system should be provided. All components should be shown on the engineering drawings in the appendix. Likewise, the confinement system for fissile material packages should be defined. The confinement system is composed of those features that are intended to ensure criticality safety, such as features that are designed to retain and provide geometry control of the fissile material.

1.2.2 Contents

This section should state the quantity of radionuclides to be transported. The description should include the following information, if appropriate:

1.   General nature of contents (e.g., irradiated fuel, metallurgical specimens, radiographic source);

2.   Identification and maximum quantity (radioactivity or mass) of the radioactive material;

3.   Identification and quantity limits of fissile material;

4.   Chemical and physical form, including density and moisture content, and the presence of any moderating constituents;

5.   Location and configuration of contents within the packaging, including secondary containers, wrapping, shoring, and other material not defined as part of the packaging;

6.   Identification and quantity of nonfissile material used as neutron absorbers or moderators;

7.   Any material subject to chemical, galvanic, or other reaction, including the generation of gases;

8.   Maximum weight of radioactive contents and maximum weight of payload, including secondary containers and packaging, if applicable;

9.   Maximum decay heat;

10.   Any loading restrictions; and

11.   For irradiated nuclear fuel:

    a)   Type of fuel and fuel assembly specifications, including the number of fuel rods and dimensional data for fuel rods and assembly structure,

    b)   Control assemblies or other contents (e.g., startup sources),

    c)   Initial fissile mass,

    d)   Maximum irradiation and minimum irradiation, if applicable,

e)   Minimum cooling time,

f)   Initial enrichment (maximum and minimum, if applicable),

g)   Unique or unusual conditions, such as damaged fuel, non-uniform enrichments, and annular pellets,

h)   Cavity fill gas, and

i)   Estimates of surface contamination.

This section should identify any contents contained in any other class of hazardous material (other than Class 7) covered in the latest edition of the United Nations, "Recommendations on the Transport of Dangerous Goods" Fourteenth Revised Edition, issued in 2005 (Ref. 5) (e.g., explosive, pyrophoric, corrosive, flammable, oxidizing). This description should include the chemical and physical form of the material, the quantity limits of the material, and how the design of the packaging accounts for the properties of such contents.

Additionally, this section should provide a description of the contents that is suitable for inclusion in the certificate, including the type and form of material and the maximum quantity of material per package.

1.2.3 Special Requirements for Plutonium

For packages that may contain plutonium in excess of 0.74 terabecquerels (TBq) (20 curies (Ci)) per package, this section should show that these contents must be in solid form, in accordance with 10 CFR 71.63, "Special Requirements for Plutonium Shipments."

> A similar requirement does not exist in Canada and, therefore, for approval in the United States, compliance with 10 CFR 71.63 must be demonstrated. The application should include only contents in solid form for plutonium in excess of 0.74 TBq (20 Ci).

1.2.4 Operational Features

In the case of a complex package system, this section should describe the operational features of the package. This should include a schematic diagram showing all valves, connections, piping, openings, seals, containment boundaries, and so forth.

1.3 Appendix

The appendix should include the engineering drawings for the packaging. The drawings should clearly detail the safety features considered in the package evaluation. Packages authorized for shipment must conform to the approved design; that is, each packaging must be fabricated in exact conformance to the drawings submitted for and referenced in the approval. While the details included in the drawings should be at a level that is sufficient for independent verification and certification purposes, it should be noted that additional details on certification drawings, similar to those needed for fabrication drawings, would require frequent review and approval by regulatory authorities if any changes are made.

Each drawing should include a title block that identifies the preparing organization, drawing number, sheet number, title, date, and signature or initials indicating approval of the drawing.

Revised drawings should identify the revision number, date, and description of the change in each revision. NUREG/CR-5502, "Engineering Drawings for 10 CFR Part 71 Package Approvals," issued May 1998 (Ref. 6), includes information that may be useful in developing and reviewing engineering drawings. The drawings should include the following:

1. General arrangement of packaging and contents, including dimensions;

2. Design features that affect the package evaluation;

3. Packaging markings (e.g., model number, serial number, gross weight, and assigned identification number);

4. Maximum allowable weight of package;

5. Maximum allowable weight of contents and secondary packaging;

6. Minimum weights, if appropriate; and

7. Materials of construction, including appropriate material specifications and a materials list.

Information on design features should include the following as appropriate:

1. Identification of the design feature and its components;

2. Codes, standards, or other similar specification documents for fabrication, assembly, and testing;

3. Location with respect to other package features;

4. Dimensions, with appropriate tolerances;

5. Operational specifications (e.g., bolt torque);

6. Weld design and inspection method; and

7. Gasketed joints in the containment system with sufficient detail to show, at a minimum, the surface finish and flatness requirements of the closure surfaces, the gasket or O-ring specification, and, if appropriate, the method of gasket or O-ring retention.

The appendix should also include a list of references, applicable pages from referenced documents that are not generally available, supporting information on special fabrication procedures, determination of the package category, and other appropriate supplemental information.

A generic sketch representing the package as prepared for transport is required in order to comply with Paragraph 807(h) of TS-R-1 as referenced in Paragraph 7(1)(a) of the PTNS Regulations. This sketch is required under the Canadian regulations. The appendix should include a generic sketch which represents the package as prepared for transport.

2. Structural Evaluation

This section of the application should identify, describe, discuss, and analyze the principal structural design of the packaging, components, and systems important to safety. In addition, this section should describe how the package complies with the performance requirements of 10 CFR Part 71 and the PTNS Regulations.

2.1 Description of Structural Design

2.1.1 Discussion

This section should identify the principal structural members and systems (such as the containment vessel, impact limiters, radiation shielding, closure devices, and ports) that are important to the safe operation of the package. The discussion should reference the locations of these items on drawings and discuss their structural design and performance.

The packaging should be described in sufficient detail to provide an adequate basis for its evaluation. The text, sketches, and data describing the structural design features should be consistent with the engineering drawings and the models used in the structural evaluation. Descriptive information important to structures includes the following:

1. Dimensions, tolerances, and materials;

2. Maximum and minimum weights and centers of gravity of packaging and major subassemblies;

3. Maximum and minimum weight of contents, if appropriate;

4. Maximum normal operating pressure;

5. Description of closure system;

6. Description of handling requirements; and

7. Fabrication methods, if appropriate.

2.1.2 Design Criteria

This section should describe the load combinations and factors that serve as design criteria. For each criterion, this section should state the maximum allowable stresses and strains (as a percentage of the yield or ultimate values for ductile failure) and describe how the other structural failure modes (e.g., brittle fracture, fatigue, buckling) are considered. If different design criteria are to be allowed in various parts of the packaging or for different conditions, this section should indicate the appropriate values for each case. This section should identify the criteria that are used for impact evaluation as well as the codes and standards that are used to determine material properties, design limits, or methods of combining loads and stresses. In the event that the design criteria deviate from those specified by standard codes, or if such codes do not cover certain components, this section should provide a detailed description and justification for the use of the design criteria used as substitutes. All assumptions made must be verified. For SNF and high-level waste packages, load combinations and design criteria are defined in NRC Regulatory Guides 7.6, "Design Criteria for the Structural Analysis of Shipping

17

Cask Containment Vessels," issued March 1978 (Ref. 7), and 7.8, "Load Combinations for the Structural Analysis of Shipping Casks for Radioactive Material," issued March 1989 (Ref. 8).

2.1.3 Weights and Centers of Gravity

This section should list the total weight of the packaging and contents and tabulate the weights of major individual subassemblies such that the sum of the parts equals the total of the package. The discussion should identify the location of the center of gravity of the package and any other centers of gravity referred to in the application. For example, the center of gravity for major subassemblies or package configurations that use different shielding configurations or components should be identified. A sketch or drawing that clearly shows the individual subassembly referred to and the reference point for locating its center of gravity should be included. In general, the discussion need not provide the calculations used to determine the centers of gravity.

2.1.4 Identification of Codes and Standards for Package Design

This section should identify the established codes and standards proposed for use in package design, fabrication, assembly, testing, maintenance, and use. An assessment of the applicability of codes and standards should be included.

This section should identify established codes and standards or justify the basis used for the package design and fabrication. The codes and standards must be appropriate for the intended purpose and must be properly applied. The code or standard should consider the quantity and hazard of the radioactive contents. In specifying a code or standard, it is important to show that the code or standard:

1.   Was developed for structures of similar design and material, if not specifically for shipping packages;

2.   Was developed for structures with similar loading conditions;

3.   Was developed for structures that have similar consequences of failure;

4.   Adequately addresses potential failure modes; and

5.   Adequately addresses margins of safety.

The American Society of Mechanical Engineers (ASME) developed a code specifically for the design and construction of the containment systems of a spent fuel cask or high-level radioactive waste transport packaging, known as the ASME Boiler and Pressure Vessel Code, Section III, Division 3, "Containment Systems and Transport Packagings for Spent Nuclear Fuel and High Level Radioactive Waste" (Ref. 9). In general, use of this code is acceptable for material specifications, design, fabrication, welding, examination, testing, inspection, and certification of containment systems for spent fuel packaging. Deviations from this code should be explicitly defined and justified for spent fuel, high-level radioactive waste packages, or other packages designed to transport large quantities of radioactive material (e.g., greater than 3,000 A_1 for special form or 3,000 A_2 for normal form material).

NUREG/CR-3854, "Fabrication Criteria for Shipping Containers," issued March 1985 (Ref. 10) and NUREG/CR-3019, "Recommended Welding Criteria for Use in the Fabrication of Shipping

Containers for Radioactive Materials," issued March 1985 (Ref. 11) provide information regarding design and fabrication criteria and appropriate codes and standards for all types of radioactive material transport packages.

2.2 Materials

This section should describe the materials of construction of the package and address the requirements in 10 CFR 71.43(d) or Paragraph 613 of TS-R-1 which is incorporated in Subsection 1(1) of the PTNS Regulations by reference to Paragraph 650 of TS-R-1.

2.2.1 Material Properties and Specifications

This section should list the material mechanical properties used in the structural evaluation. These should include yield stress, ultimate stress, modulus of elasticity, ultimate strain, Poisson's ratio, density, and coefficient of thermal expansion. If impact limiters are used, this section should include either a compression stress-strain curve for the material or the force-deformation relationship for the limiter, as appropriate. For materials that are subjected to elevated temperatures, the appropriate mechanical properties under those conditions should be specified. The source of the information in this section should be identified by publication and page number. Where material properties are determined by testing, this section should describe the test procedures, conditions, and measurements in sufficient detail to enable the staff to evaluate the validity of the results. Materials of construction for fracture-critical components, including, but not limited to, containment vessels, should be resistant to brittle fracture at all design temperatures taking into account other factors which may affect performance, such as material thickness. Fracture-critical components include those components whose failure could result in release of radioactive material.

An appropriate specification should be identified for the control of each material. Materials and their properties should be consistent with the design code or standard selected. If no code or standard is available, the application should provide adequately documented material properties and specifications for the design and fabrication of the packaging.

The materials of fracture-critical structural components should have sufficient fracture toughness to preclude brittle fracture under normal conditions of transport and hypothetical accident conditions. NRC Regulatory Guides 7.11, "Fracture Toughness Criteria of Base Material for Ferritic Steel Shipping Cask Containment Vessels with a Maximum Wall Thickness of 4 Inches (0.1 m)," issued June 1991 (Ref. 12), and 7.12, "Fracture Toughness Criteria of Base Material for Ferritic Steel Shipping Cask Containment Vessels with a Wall Thickness Greater than 4 Inches (0.1 m) But Not Exceeding 12 Inches (0.3 m)," issued June 1991 (Ref. 13), provide criteria for fracture toughness.

The material properties should be appropriate for the load conditions (e.g., static or dynamic impact loading, hot or cold temperatures, and wet or dry conditions). The temperatures at which allowable stress limits are defined should be consistent with minimum and maximum service temperatures. Force-deformation properties for impact limiters should be based on appropriate test conditions and temperatures.

For packages with impact limiting devices or features, the methods used to establish their force-deflection characteristics should be provided and should include testing. Testing of the impact limiters may be carried out statically if the effect of strain rates on the material crush properties is accounted for and properly included in the force-deflection relationship for impact analysis.

The force-deflection curve of the impact limiter should be provided for all directions analyzed for the packaging.

2.2.2 Chemical, Galvanic, or Other Reactions

This section should describe possible chemical, galvanic, or other reactions in the packaging or between the packaging and the package contents, as well as methods used to prevent significant reactions. For each component material of the packaging, this section should list all chemically or galvanically dissimilar materials in contact with it. Coatings used on internal or external package surfaces, any reactions resulting from water in-leakage or cask flooding, and the possible generation of hydrogen or other gases from chemical, radiolytic, or other interactions should be considered. Galvanic interactions and the formation of a eutectic for components that are or may be in physical contact should also be considered. Such interactions may occur with depleted uranium, lead, or aluminum in contact with steel. If appropriate, the application should consider the embrittling effects of hydrogen, taking into account the metallurgical state of the packaging materials. Pyrophoricity should also be addressed.

2.2.3 Effects of Radiation on Materials

This section should describe any aging or damaging effects of radiation on the packaging materials and should cite references for established dose limits for affected materials. These should include degradation of seals, sealing materials, coatings, adhesives, and structural materials.

2.3 Fabrication and Examination

2.3.1 Fabrication

This section should describe the fabrication processes used for the package, such as fitting, aligning, welding and brazing, heat treatment, and foam and lead pouring. For fabrication specifications prescribed by an acceptable code or standard (e.g., those promulgated by ASME or the American Welding Society), the engineering drawings should clearly specify the code or standard, edition, date, or addenda. Unless the application justifies otherwise, specifications of the same code or standard used for design should also be used for fabrication. For components for which no code or standard is applicable, the application should identify the specifications on which the evaluation depends and describe the method of control to ensure that these specifications are achieved. This description should reference quality assurance or other appropriate specifications documents, which should be identified on the engineering drawings.

2.3.2 Examination

This section should describe the methods and criteria by which the fabrication is determined to be acceptable. Unless the application justifies otherwise, specifications of the same code or standard used for fabrication should also be used for examination. For components for which no fabrication code or standard is applicable, the application should summarize the examination methods and acceptance criteria in Section 8, "Acceptance Tests and Maintenance Program."

2.4 General Requirements for All Packages

This section should address the requirements of 10 CFR 71.43(a), (b), and (c) or Paragraphs 634, 635, and 639 of TS-R-1 which are incorporated in Subsection 1(1) of the PTNS Regulations by reference to Paragraph 650 of TS-R-1.

2.4.1 Minimum Package Size

This section should specify the smallest overall dimension of the package, which should not be less than 10 cm (4 in.).

2.4.2 Tamper-Indicating Feature

This section should describe the package closure system in sufficient detail to show that it incorporates a protective feature that, while intact, is evidence that unauthorized persons have not tampered with the package. The description should include covers, ports, or any other access that must be closed during normal transportation. Tamper indicators and their locations should be described.

2.4.3 Positive Closure

This section should describe the package closure system in sufficient detail to show that it cannot be inadvertently opened. This description should include covers, valves, or any other access that must be closed during normal transportation.

2.5 Lifting and Tie-down Standards for All Packages

2.5.1 Lifting Devices

This section should identify all devices and attachments that can be used to lift the package or its lid and show by testing or analysis that these devices comply with the requirements of 10 CFR 71.45(a) or Paragraphs 607 and 608 of TS-R-1 which are incorporated in Subsection 1(1) of the PTNS Regulations by reference to Paragraph 650 of TS-R-1. This includes demonstrating that failure of the lifting devices under excessive load will not impair the ability of the package to meet other requirements. This section should also include drawings or sketches that show the locations and construction of these devices and should show the effects of the forces imposed by lifting devices on other packaging surfaces. Documented values of the yield stresses of the materials should be used as the criteria for demonstrating compliance with 10 CFR 71.45(a), including failure under excessive load. For attachments or other features that could be used to lift the package and that do not meet the lifting criterion, this section should show how they are rendered inoperable for lifting.

Canadian regulations do not specify a numerical criterion for acceleration load factors or snatch factors for lifting fixtures that are a structural part of the package as per Paragraphs 607 and 608 of TS-R-1. The criterion specified in 10 CFR 71.45(a) is a minimum safety factor of 3 against yielding. This section should show that the lifting devices meet the criterion in 10 CFR 71.45(a).

2.5.2 *Tie-down Devices*

This section should describe the overall tie-down system for the package and show that the system meets the requirements of 10 CFR 71.45(b) and Paragraph 636 of TS-R-1 which is incorporated in Subsection 1(1) of the PTNS Regulations by reference to Paragraph 650 of TS-R-1. Any device that is a structural part of the package and can be used for tie-down should be identified. Drawings or sketches that show the locations and construction of the overall tie-down system and the individual devices should be provided. This section should also discuss the testing or analysis that shows that these devices are designed to withstand tie-down forces and should show the effect of the imposed forces on vital package components, including the interfaces between the tie-down devices and other package surfaces. Documented values of the yield stresses of the materials should be used as the criteria for demonstrating the adequacy of the tie-down devices and failure under excessive load. This section should show that failure of the tie-down devices under excessive load will not impair the ability of the package to meet other requirements.

Canadian regulations do not specify numerical design criteria for tie-down devices as per Paragraph 636 in TS-R-1. The design criteria for tie-down devices that are a structural part of the package are defined in 10 CFR 71.45(b) as follows:

> The system must be capable of withstanding, without generating stress in any material of the package in excess of its yield strength, a static force applied to the center of gravity of the package having a vertical component of 2 times the weight of the package with its contents, a horizontal component along the direction in which the vehicle travels of 10 times the weight of the package with its contents, and a horizontal component in the transverse direction of 5 times the weight of the package with its contents.

This section should show that the tie-down devices meet the criteria of 10 CFR 71.45(b). For attachments or other features that are a structural part of the package, that could be used for tie-down and that do not meet the tie-down criteria, this section should show how they are rendered inoperable for tie-down.

2.6 Normal Conditions of Transport

This section should describe the evaluation that shows that the package meets the standards specified in 10 CFR 71.43(f) and 10 CFR 71.51(a)(1) or Paragraphs 646 and 656(a) of TS-R-1 which are incorporated in Subsection 1(1) of the PTNS Regulations by reference to Paragraph 650 of TS-R-1 when subjected to the tests and conditions specified in 10 CFR 71.71, "Normal Conditions of Transport," or Paragraphs 719–724 of TS-R-1 which are incorporated in Subsection 1(4) of the PTNS Regulations by reference to Paragraph 716 of TS-R-1 (normal conditions of transport). The package should be evaluated against each condition individually. The evaluation should show that the package satisfies the applicable performance requirements specified in the regulations (e.g., there should be no loss or dispersal of contents; no structural changes that reduce the effectiveness of components required for shielding, heat transfer, criticality control, or containment; and no changes that would affect the ability of the package to withstand the accident conditions tests).

The structural evaluation of the package under normal conditions of transport may be performed by analysis or test or a combination of both. In describing the structural evaluation of the package, this section should clearly show that the most limiting initial test conditions and most damaging orientations have been considered and the evaluation methods are appropriate and properly applied. An adequate set of test orientations should be considered since the most damaging orientation for one component may not be the most damaging for another component. The evaluation methods should be appropriate for the loading conditions considered and follow accepted practices and precepts. The results should be correctly interpreted.

In addressing the sections listed below, the following general information should be considered and included as appropriate:

1.   For evaluation by test, this section should describe the test method, procedures, equipment, and facilities that were used. For example, the drop test surface should be described sufficiently to show that it represents an essentially unyielding surface. The steel puncture bar should be described with respect to its material of construction; dimensions, including showing that the length is sufficient to cause maximum damage to the package; and the method used to secure the bar to the unyielding surface. The test methods and instruments should be adequate for the measurements needed, and the measurements should be sufficient for describing the structural response or damage. The pass/fail criteria for evaluating the package performance in the tests should be provided and justified.

2.   The package orientations evaluated for the tests should be clearly identified and justified as being most damaging. Where sequential tests are required, the damage from one test should be considered when performing subsequent tests.

3.   If the package tested is not identical in all respects to the package described in the application, the differences should be identified, and justification should be given to show that the differences would not affect the test results.

4.   The materials used as substitutes for the radioactive contents during the tests should be described, and justification should be given that shows that this substitution would not affect the results, including an assessment of the effects of internal decay heat and pressure buildup, if appropriate.

5.   A detailed and quantitative description of the damage caused by the tests should be provided along with the results of any measurements that were made, including both interior and exterior damage as well as photographs of the damaged packaging. Videos of the tests should be provided if available. The test results should be thoroughly evaluated. The test conclusions should be valid and defensible. Unexpected or unexplainable test results, indicating possible testing problems or non-reproducible specimen behavior, should be discussed and evaluated. The tests should demonstrate an adequate margin of safety. The test results should clearly show that the effects of the tests can be reliably reproduced. Effects of uncertainties in mechanical properties, test conditions, and diagnostics should be described.

6.   For prototype and model testing, this section should provide a complete description of the test specimen, including detailed drawings that show its dimensions and materials of construction and dimensional tolerances to which the prototype or model was fabricated. The fabrication tolerances of the test specimen should be compared to those that will be used for the package. The test specimen should be fabricated using the same

materials, methods, and quality assurance as specified in the design. For scale models, this section should identify the scale factor that was used and should provide a detailed description of the laws of similitude that were used for testing, considering time scale, material density, velocity at impact, and kinetic energy. Information should be provided to show that the model test will give conservative results for peak g-force, maximum deformation, and dissipated energy. In addition, the damage done to the model should be correlated to the damage to the package. In general, scale models do not provide reliable quantitative data regarding the leakage rate of the package.

7. For evaluation by analysis, this section should describe the methods and calculations used in the package evaluation in sufficient detail to enable the staff to verify the results. In so doing, this section should clearly describe and justify all assumptions used in the analysis and include adequate narration, sketches, and free body force diagrams. In addition, for equations used in the analysis, this section should either cite the source or include the derivation.

8. The computer programs should be identified and described and should be shown to be well benchmarked, widely used for structural analyses, and applicable to the evaluation.

9. Computer models and related details should be well described and justified. For example, the number of discrete finite elements used in the model should reflect the type of analysis performed and should be appropriate, considering such factors as stress or displacement.

10. Sensitivity studies used to determine the appropriate number of nodes or elements for a particular model should be provided.

11. A detailed description of the modeling of bolted connections, including element types, modeling technique, and material properties, should be included.

12. For impact analysis, information should be provided that shows how all of the kinetic energy will be dissipated and what local deformation and dynamic forces would occur during impact, the package response in terms of stress and strain to components and structural members, the structural stability of individual members, stresses attributable to impact combined with those stresses caused by temperature gradients, differential thermal expansions, pressure, and other loads. Load combinations and acceptance criteria are provided in References 7 and 8. The evaluation should compare the maximum stresses or strains with design code allowables. The analysis should provide information that shows that the critical combinations of environmental and loading conditions were evaluated.

13. The analytical results should be directly compared with the acceptance criteria. The response of the package to loads, in terms of stress and strain to components and structural members, should be shown. The structural stability of individual members should be evaluated, as applicable.

14. Analytical methods should consider impact at any angle, rigid-body rotation, and secondary impact (slapdown). Dynamic amplification factors should be appropriately applied if a quasi-static analysis technique has been used.

15.   The models and material properties should be appropriate for the load combinations considered, and the evaluation should include all appropriate initial conditions and load combinations. Material properties (e.g., elastic, plastic) should be consistent with the analysis methods. The strain rate at which the properties were determined should be justified. The analysis should consider true stress-strain or engineering stress-strain, as applicable.

16.   An assessment should be included that shows that the normal conditions do not reduce the effectiveness of the package.

2.6.1 Heat

The thermal evaluation for the heat test should be described and reported in Section 3, "Thermal Evaluation." The results of the thermal evaluation should be used as input to the following sections.

2.6.1.1 Summary of Pressures and Temperatures

This section should summarize all pressures and temperatures derived in Section 3, "Thermal Evaluation" that will be used to perform the calculations needed for Sections 2.6.1.2–2.6.1.4, as described below.

2.6.1.2 Differential Thermal Expansion

This section should present calculations of the circumferential and axial deformations and stresses (if any) that result from differential thermal expansion. The evaluation should consider possible interferences resulting from a reduction in gap sizes. Steady-state and transient conditions should be considered. These calculations should be sufficiently comprehensive to demonstrate package integrity under normal transport conditions and should consider appropriate load combinations, such as maximum normal operating pressure and decay heat and fabrication stresses.

2.6.1.3 Stress Calculations

This section should present calculations of the stresses that are attributable to the combined effects of thermal gradients, pressure, and mechanical loads (including fabrication stresses from lead pour and lead cooldown). Sketches that show the configuration and dimensions of the members or systems being analyzed and the points at which the stresses are calculated should be provided. The analysis should consider whether repeated cycles of thermal loadings, together with other loadings, will cause fatigue failure or extensive accumulations of deformation.

2.6.1.4 Comparison with Allowable Stresses

This section should present the appropriate stress combinations and compare the resulting stresses with the design criteria specified in the application and should show that all relevant performance requirements have been satisfied as specified in the regulations. Stresses should be within the limits for normal condition loads.

2.6.2 Cold

The thermal evaluation under normal cold conditions should be described and reported in accordance with Section 3, "Thermal Evaluation." Using the results from the thermal evaluation, this section should assess the effects that the cold condition has on the package, including material properties and possible liquid freezing and lead shrinkage. The resulting temperatures and their effects on package components and operation of the package should be reported. Brittle fracture should be evaluated. The evaluation should consider the minimum internal pressure with the minimum internal heat load (typically assumed to be no decay heat) and any residual fabrication stresses. Differential thermal expansions and possible geometric interferences should be considered. Stresses should be within the limits for normal condition loads.

2.6.3 Reduced External Pressure

This section should describe the evaluation of the package for the effects of reduced external pressure, as specified in 10 CFR 71.71(c)(3) and Paragraphs 643 and 619 of TS-R-1 which are incorporated in Subsection 1(1) of the PTNS Regulations by reference to Paragraph 650 of TS-R-1. The evaluation should include the greatest pressure difference between the inside and outside of the package, as well as the inside and outside of the containment system, and evaluate this condition in combination with the maximum normal operating pressure.

> There are some differences in the provisions of 10 CFR 71.71(c) and Paragraphs 643 and 619 of TS-R-1. Paragraph 643 of TS-R-1 specifies a reduced ambient pressure of 60 kilopascals (kPa) (8.7 pound-force per square inch ($lbf/in.^2$)), and 10 CFR 71.71(c) specifies a reduced ambient pressure of 25 kPa (3.5 $lbf/in.^{2)}$ absolute. Paragraph 619 of TS-R-1 specifies the reduced ambient pressure for air transport. This section should show that the package meets all three requirements, unless the package will not be transported by air, in which case Paragraph 619 does not apply.

2.6.4 Increased External Pressure

This section should describe the evaluation of the package for the effects of increased external pressure, as specified in 10 CFR 71.71(c)(4) and Paragraph 615 of TS-R-1 which is incorporated in Subsection 1(1) of the PTNS Regulations by reference to Paragraph 650 of TS-R-1. The evaluation should include the greatest pressure difference between the inside and outside of the package, as well as the inside and outside of the containment system, and evaluate this condition in combination with the minimum internal pressure. This section should include a buckling evaluation.

> Since 10 CFR 71.71(c)(4) includes a specific value for the increased external pressure and there is no analogous value in TS-R-1, this section should show that the package can withstand the increased external pressure defined in 10 CFR 71.71(c)(4) (i.e., 140 kPa (20 $lbf/in.^2$) absolute).

2.6.5 Vibration

This section should describe the evaluation of the package for the effects of vibrations that are normally incident to transport, as specified in 10 CFR 71.71(c)(5) or Paragraph 612 of TS-R-1 which is incorporated in Subsection 1(1) of the PTNS Regulations by reference to Paragraph 650 of TS-R-1. The combined stresses attributable to vibration, temperature, and pressure loads should be considered, and a fatigue analysis should be included if applicable. If closure bolts are reused, the bolt preload should be considered in the fatigue evaluation. Packaging components, including internals, should be evaluated for resonant vibration conditions that can cause rapid fatigue damage.

2.6.6 Water Spray

This section should show that the water spray test has no significant effect on the package.

2.6.7 Free Drop

This section should describe the package evaluation for the effects of a free drop. The general comments in Section 2.7.1 below may also apply to this condition. Note that the free-drop test follows the water spray test. This section should also address such factors as drop orientation; effects of free drop in combination with pressure, heat, and cold temperatures; and other factors discussed in Section 2.6 of this document.

Closure lid bolts should be evaluated for the combined effects of free-drop impact force, internal pressures, thermal stress, O-ring compression force, and bolt preload. Port covers, port cover plates, and shielding enclosures should also be evaluated for the combined effects.

2.6.8 Corner Drop

If applicable, this section should describe the effects of corner drops on the package. The applicability of the corner drop is defined in 10 CFR 71.71(c)(8) and Paragraph 722 of TS-R-1 which is incorporated in Subsection 1(4) of the PTNS Regulations by reference to Paragraph 716 of TS-R-1.

2.6.9 Compression or Stacking

This section should describe the effects of the compression or stacking test. The package must be subjected for a period of 24 hours to a compressive load equal to the greater of the following:

1. The equivalent of 5 times the weight of the package;

2. The equivalent of 13 kPa (2 lbf/in.^{2}) multiplied by the vertically projected area of the package.

The load shall be applied uniformly to the top and bottom of the package in the position in which the package would normally be transported.

Canadian regulations specify that the compression (or stacking) test does not need to be considered if the shape of the packaging effectively prevents stacking, as stated in Paragraph 723 of TS-R-1 which is incorporated in Subsection 1(4) of the PTNS Regulations by reference to Paragraph 716 of TS-R-1. U.S. regulations in 10 CFR 71.71(c)(9) do not include the exception; however, the compression test is only required for small, light packages with a mass less than 500 kilograms (kg) (1,100 pounds (lb)). For this section, the evaluation should consider the effects of the compression test for the following:

1. All packages with mass less than 500 kg (1,100 lb); and

2. Packages with mass greater than 500 kg (1,100 lb) if the shape of the packaging does not prevent stacking.

2.6.10 Penetration

This section should describe the effects of penetration on the package and should identify the most vulnerable location on the package surface.

2.7 Hypothetical Accident Conditions

This section should describe the structural performance of the package when subjected to the tests specified at 10 CFR 71.73, "Hypothetical Accident Conditions," or the tests described in Paragraphs 726–729 of TS-R-1 which are incorporated in Subsection 1(4) of the PTNS Regulations by reference to Paragraph 716 of TS-R-1.

The structural evaluation should consider the accident conditions in the indicated sequence to determine their cumulative effect on a package. Damage caused by each test is cumulative, and the evaluation of the ability of a package to withstand any one test must consider the damage that resulted from the previous tests. This section should confirm that the package effectiveness has not been reduced as a result of the normal conditions of transport, as included in Section 2.6 above. Brittle fracture should also be considered. This section should include applicable information regarding tests and analyses, as described in Section 2.6 of this document. In general, inelastic deformation of the containment system closure (e.g., bolts, flanges, seal regions) is not acceptable for Type B packages. Deformation of other parts of the containment vessel may be acceptable if the containment boundary is not compromised. Deformation of shielding components, components required for heat transfer and insulation, and components required for subcriticality should be defined and evaluated in Sections 3, "Thermal Evaluation," 4, "Containment," 5, "Shielding Evaluation," and 6, "Criticality Evaluation" of the application for package approval.

With respect to initial conditions for the tests (except for the water immersion tests), ambient temperature and internal pressure should be specified and should be shown to be the most unfavorable. For physical tests that are not performed at the most unfavorable pressure or at the temperature extremes, the application should include an evaluation to show that the pressure and temperature would not affect the ability of the package to meet the other performance requirements. For example, the evaluation may include information regarding combined loads and material properties.

Paragraph 664 of TS-R-1 which is incorporated in Subsection 1(1) of the PTNS Regulations by reference to Paragraph 650 of TS-R-1 requires that a Type B package be designed for an ambient temperature range from −40° C (−40° F) to +38° C (+100° F). Initial conditions in 10 CFR 71.73(b) specify that the ambient temperature preceding and following the tests must be between −29° C (−20° F) and +38° C (+100° F).

The cold temperature condition that is to be considered as the initial condition for the accident tests is different for U.S. and Canadian regulations. Paragraph 664 of TS-R-1 specifies −40° C (−40° F) and 10 CFR 71.73(b) specifies −29° C (−20° F). Therefore, the cold temperature considered as the initial condition for the hypothetical accident conditions drop test must be −40° C (−40° F).

Paragraph 727 of TS-R-1 which is incorporated in Subsection 1(4) of the PTNS Regulations by reference to Paragraph 716 of TS-R-1 specifies that the accident condition drops (drop I, the 9 meter (m) (30 ft) free drop; drop II, the puncture test; and drop III, the crush test) be performed, as applicable, in whichever order results in the maximum damage, considering the subsequent application of the fire test. In addition, no package is required to be subjected to both the 9 m (30 ft) free drop and the crush test. On the other hand, 10 CFR 71.73 specifies that the sequence of the tests must first be the 9 m (30 ft) free drop, followed by the crush test for certain packages, and then followed by the puncture test. In this section, the application should specifically address the most restrictive conditions. If the 9 m (30 ft) drop is performed first, the application must include a justification that this sequence results in maximum damage, also considering the subsequent fire test. If there is evidence that performing the puncture test before the 9 m (30 ft) drop will result in maximum damage, then two puncture tests should be performed, one prior to the 9 m drop (30 ft) and one following the 9 m (30 ft) drop. For packages requiring the crush test, the accident sequence must include a 9 m (30 ft) drop, followed by the crush test.

2.7.1 Free Drop

This section should evaluate the package under the free-drop test. The performance and structural integrity of the package should be evaluated for the drop orientation that causes the most severe damage, including center-of-gravity-over-corner orientation, oblique orientation with secondary impact (slap down), side drop, and drop onto the closure. Orientations for which the center of gravity is directly over the point of impact should also be considered. An orientation that results in the most damage to one system or component may not be the most damaging for other systems and components. If a feature such as a tie-down component is a structural part of the package, it should be considered in selecting the drop test configurations and drop orientation. For these reasons, it is usually necessary to consider several drop orientations.

The following items should be addressed, if applicable:

1. For packages with lead shielding, the package should be evaluated for the effects of lead slump. The lead slump determined should be consistent with that used in the shielding evaluation.

2. The closure lid bolt design should be assessed for the combined effects of free drop impact force, internal pressures, thermal stress, O-ring compression force, and bolt preload.

3. The buckling of package components should be evaluated.

4.   Other package components, such as port covers, port cover plates, and shield enclosures, should be evaluated for the combined effects of package drop impact force, puncture, internal pressures, and thermal stress.

2.7.1.1 End Drop

This section should describe the effects of the end-drop test on the package.

2.7.1.2 Side Drop

This section should describe the effects of the side-drop test on the package.

2.7.1.3 Corner Drop

This section should describe the effects of the corner-drop test on the package.

2.7.1.4 Oblique Drops

This section should describe the effects of oblique drops or should provide information that shows that the end, side, and corner drops are more damaging to all systems and components that are vital to safety.

2.7.1.5 Summary of Results

This section should describe the condition of the package after each drop test and describe the damage for each orientation.

2.7.2 Crush

If applicable, this section should describe the effects of the dynamic crush test on the package.

Canadian regulations require that the crush test (drop III) be substituted for the 9 m (30 ft) drop (drop I) for certain packages. U.S. regulations require that both tests be performed (the 9 m (30 ft) drop followed by the crush test) for these packages. The package type that is subjected to the crush test is the same in both regulations and is based on the weight, density, and authorized contents of the package. For packages requiring the crush test, the accident sequence must include a 9 m (30 ft) drop, followed by the crush test.

2.7.3 Puncture

This section should describe the effects of puncture on the package and identify and justify that the orientations for which maximum damage would be expected have been evaluated. This description should consider any damage resulting from the free-drop and crush tests, as well as both local damage near the point of impact of the puncture bar and the overall effect on the package. Containment system valves and fittings should be addressed. Punctures at oblique angles, near a support valve, at the package closure, and at a penetration should be considered as appropriate. General comments provided in Sections 2.6 and 2.7.1 of this document may also apply to this test condition.

Although analytical methods are available for predicting puncture, empirical formulas derived from puncture test results of laminated panels are usually used for package design. The Nelm's formula, developed specifically for package design, provides the minimum thickness needed for preventing the puncture of the steel surface layer of a typical steel-lead-steel laminated cask wall.

2.7.4 Thermal

The thermal test should follow the free drop and puncture tests and should be reported in Section 3, "Thermal Evaluation," of the application. This section should evaluate the structural design for the effects of a fully engulfing fire as specified in 10 CFR 71.73(c)(4) or Paragraph 728 of TS-R-1 which is incorporated in Subsection 1(4) of the PTNS Regulations by reference to Paragraph 716 of TS-R-1. Any damage resulting from the free drop, crush, and puncture conditions should be incorporated into the initial condition of the package for the fire test. The temperatures resulting from the fire and any increase in gas inventory caused by combustion or decomposition processes should be considered when determining the maximum pressure in the package during or after the test. The maximum thermal stresses that can occur either during or after the fire should be addressed.

2.7.4.1 Summary of Pressures and Temperatures

This section should summarize all of the temperatures and pressures as determined in Section 3, "Thermal Evaluation," of the application.

2.7.4.2 Differential Thermal Expansion

This section should include calculations of the circumferential and axial deformations and stresses (if any) that result from differential thermal expansion. Peak conditions, post-fire steady-state conditions, and all transient conditions should be considered.

2.7.4.3 Stress Calculations

This section should include calculations of the stresses caused by thermal gradients, differential expansion, pressure, and other mechanical loads. Sketches showing configuration and dimensions of the members of systems under investigation and locations of the points at which the stresses are being calculated should be included.

2.7.4.4 Comparison with Allowable Stresses

This section should make the appropriate stress combinations and compare the resulting stresses with the design criteria as described in Section 2.1.2 of the application. This section should show that all the performance requirements specified in the regulations have been satisfied.

2.7.5 Immersion—Fissile Material

If the contents include fissile material subject to the requirements of 10 CFR 71.55, "General Requirements for Fissile Material Packages," or Paragraph 671 of TS-R-1 which is incorporated in Paragraph 7(1)(a) of the PTNS Regulations by reference to Paragraph 813 of TS-R-1 (unless excepted by Paragraph 671), and if water leakage has not been assumed for the criticality analysis, this section should assess the effects and consequences of the water immersion test condition in accordance with 10 CFR 71.73(c)(5) or Paragraphs 731–733 of TS-R-1 which are incorporated in Subsection 1(4) of the PTNS Regulations by reference to Paragraph 716 of TS-R-1. The test should consider immersion of a damaged specimen under a head of water of at least 0.9 m (3 ft) in the orientation for which maximum leakage is expected.

2.7.6 Immersion—All Packages

This section should evaluate, as required in 10 CFR 71.73(c)(6) or Paragraph 729 of TS-R-1 which is incorporated in Subsection 1(4) of the PTNS Regulations by reference to Paragraph 716 of TS-R-1, an undamaged package for water pressure equivalent to immersion under a head of water of at least 15 m (50 ft). Paragraph 729 of TS-R-1 specifies that the test period be not less than 8 hours, whereas 10 CFR 71.73(c)(6) does not specify a test duration. For test purposes, an external water pressure of 150 kPa (21.7 lbf/in.^{2}) gauge is considered to meet these conditions.

> The immersion test should be evaluated for a period of not less than 8 hours, as specified in Paragraph 729 of TS-R-1.

2.7.7 Deep Water Immersion Test (for Type B Packages Containing More than 10^5 A$_2$)

If applicable, this section should evaluate the package for an external water pressure of 2 megapascals (MPa) (290 lbf/in.^{2}) for a period of no less than 1 hour, as specified in 10 CFR 71.61, "Special Requirements for Type B Packages Containing more than 10^5 A$_2$," or Paragraphs 670 of TS-R-1, which is incorporated in Subsection 1(1) of the PTNS Regulations by reference to Paragraph 667 of TS-R-1, and 730 of TS-R-1, which is incorporated in Subsection 1(4) of the PTNS Regulations by reference to Paragraph 716 of TS-R-1. The NRC regulations in 10 CFR 71.61 state, "A Type B package containing more than 10^5 A$_2$ must be designed so that its undamaged containment system can withstand an external water pressure of 2 MPa (290 lbf/in.^{2}) for a period of not less than 1 hour without collapse, buckling, or in-leakage of water."

Paragraph 730 of TS-R-1 states, for the enhanced water immersion test: "The specimen shall be immersed under a head of water of at least 200 m for a period of not less than one hour. For demonstration purposes, an external gauge pressure of at least 2 MPa shall be considered to meet these conditions."

> These regulations differ primarily in the application of the external pressure (TS-R-1 states "immersion of the specimen," and 10 CFR 71.61 states "undamaged containment system") and in the acceptance standard (TS-R-1 states "no rupture" and 10 CFR 71.61 states "without collapse, buckling, or in-leakage of water"). The NRC requirements are more restrictive since application of the pressure is on the containment system and the acceptance standards are more restrictive (i.e., in-leakage of water is acceptable under TS-R-1 standard but not under the 10 CFR 71.61 standard). Therefore, this section should show that the package meets the most restrictive standard defined in 10 CFR 71.61.

2.7.8 Summary of Damage

This section should summarize the condition of the package after the accident test sequence. The description should address the extent to which safety systems and components have been damaged and relate the package condition to the acceptance standards.

2.8 Accident Conditions for Air Transport of Plutonium or Packages with Large Quantities of Radioactivity

This section should show that the contents of the package when transported by air will be limited such that both NRC and CNSC regulations are met. This section should specifically address the following limits:

1. Packages containing radioactive material in special form that exceeds 3,000 A_1 or 100,000 A_2 may not be transported by air.

2. Packages containing radioactive material in normal form that exceeds 3,000 A_2 may not be transported by air.

3. Packages that contain plutonium in excess of an A_2 quantity (except for material with very low concentration) may not be transported by air.

2.9 Accident Conditions for Fissile Material Packages for Air Transport

If applicable, this section should address the hypothetical accident conditions specified in 10 CFR 71.55(f) or Paragraph 680 of TS-R-1 which is incorporated in Subsection 1(1) of the PTNS Regulations by reference to Paragraph 672 of TS-R-1.

2.10 Special Form

For packages designed to transport radioactive material only in special form, this section should state that the contents meet the requirements in 10 CFR 71.75, "Qualification of Special Form Radioactive Material," or Paragraph 603 of TS-R-1 as referenced in Subsection 1(1) of the PTNS Regulations when subjected to the applicable test conditions of 10 CFR 71.75 or Paragraphs 704–711 of TS-R-1 which are incorporated in Subsection 1(1) of the PTNS Regulations by reference to Paragraphs 602-604 of TS-R-1. The chemical and physical form should be specified. In addition, this section should include a detailed drawing of the encapsulation showing its dimensions, materials, manner of construction, and method of nondestructive examination.

For approval in Canada, the application should also include a copy of the special form certificate and drawings for each capsule authorized in the package. Provision for allowance of similar capsules meeting the requirements for special form radioactive material may be granted, provided that the application includes proper demonstration that these would be bounded by the analysis provided.

2.11 Fuel Rods

The structural integrity of fuel rods and cladding integrity should be addressed for packages used to transport fresh or irradiated nuclear fuel. Where fuel structural components and cladding are considered to provide containment of radioactive material or confinement or geometry control of fissile material under normal or accident test conditions, this section should provide an analysis or test results showing that the components will maintain sufficient mechanical integrity to provide the degree of containment or confinement assumed.

For SNF, the application should specifically address whether damaged or high-burnup fuel is to be transported. High-burnup fuel for light-water reactors is defined as fuel with greater than 45,000 megawatt-days per metric ton uranium (MWD/MTU) burnup. Damaged fuel should be defined and assessed with respect to the containment, shielding, and criticality evaluations. ANSI N14.33, "Characterizing Damaged Spent Nuclear Fuel for the Purpose of Storage and Transport" (Ref. 14) provides guidance with respect to the definition of damaged fuel. Damage may include known or suspected cladding defects, greater than hairline cracks or pinhole leaks, or damage to the structural components of a fuel assembly, such as spacer grids. Any special provisions for transporting damaged fuel (e.g., canister use) should be addressed.

2.12 Appendix

The appendix should include a list of references, including chapter, section, or page numbers if appropriate; applicable pages from referenced documents if not generally available; computer code descriptions; input and output files; test results; test reports; descriptions of test facilities; and instrumentation, photographs, and other appropriate supplemental information. This appendix should also include materials and manufacturing specifications for items that are significant with respect to safety but are not produced to generally recognized standards.

3. Thermal Evaluation

This section of the application should identify, describe, discuss, and analyze the principal thermal engineering design of the packaging, components, and systems that are important to safety and describe how the package complies with the performance requirements of 10 CFR 71.33(b)(5), 10 CFR 71.33(b)(7), 10 CFR 71.43(d), 10 CFR 71.43(g), 10 CFR 71.55(f)(1)(iv), 10 CFR 71.71(c)(1) and (2), and 10 CFR 71.73(c)(4) or Paragraphs 642, 651, 652, 653, 654, 655, 660, 661, and 662 of TS-R-1, which are incorporated in Subsection 1(1) of the PTNS Regulations by reference to Paragraph 650 of TS-R-1, and Paragraphs 728 and 736 of TS-R-1 which are incorporated in Subsection 1(4) of the PTNS Regulations by reference to Paragraph 716 of TS-R-1.

This section should address the thermal performance of the package under normal and hypothetical accident conditions of transport in terms of the maximum allowable temperatures and subsequent effect on the containment, structural, shielding, and criticality safety design. Any operational, fabrication and maintenance requirements with respect to the thermal evaluation important for the package safety should be included in Sections 7, "Package Operations," and 8, "Acceptance Tests and Maintenance Program," of the application.

3.1 Description of Thermal Design

This section should describe the significant thermal design features and operating characteristics of the package and discuss the operation of all subsystems. The thermal criteria that will be directly applied to thermal results (e.g., maximum fuel temperature, shield temperature not to exceed melt) should be identified. Properties evaluated in this section, but used to support other evaluations (e.g., pressure, temperature, distributions relative to thermal stress) should also be identified. The significant results of the thermal analysis or tests and the implication of these results on the overall thermal performance of the package should be summarized. The minimum and maximum decay heat loads assumed in the thermal evaluation should be specified. The maximum decay heat load assumed should include the energy from all source terms contained in the package, including those that might be neglected in the shielding and containment analyses.

3.1.1 Design Features

This section should describe the design features of the package that are important for thermal performance. The package design must not rely on mechanical cooling systems to meet containment requirements specified in Section 4, "Containment," of this document.

Packaging design features important for the thermal evaluation include the following:

1. Package geometry and materials of construction;

2. Structural and mechanical features that may affect heat transfer, such as cooling fins, insulating materials, surface conditions of the package components, and gaps or physical contacts between internal components; and

3. The identity and volume of any coolants, if applicable.

3.1.2 Decay Heat of Contents

The maximum decay heat and the radioactivity of the contents should be specified in accordance with the requirements of 10 CFR 71.33(b)(7) or Paragraph 651 of TS-R-1 as incorporated in Subsection 1(1) of the PTNS Regulations by reference to Paragraph 650 of TS-R-1. This section should show that the derivation of the decay heat is consistent with the maximum quantity of radioactive contents. In the case of SNF packages, the computer codes discussed in Section 5, "Shielding Evaluation," of this document for determination of neutron and gamma sources may be used for calculating content decay heat loads.

3.1.3 Summary Tables of Temperatures

This section should present summary tables of the maximum or minimum temperatures that affect structural integrity, containment, shielding, and criticality under both normal conditions of transport and hypothetical accident conditions. All information presented in this table should be consistent with the information used within other sections of the application. For the fire test condition, the tables should include the following information:

1.   Maximum temperatures of various package components and the time at which they occur after fire initiation (e.g., containment vessel, seals, shielding, fuel/cladding); and

2.   Temperatures of the post-fire steady-state condition.

3.1.4 Summary Tables of Maximum Pressures

The summary tables should include the maximum normal operating pressure and maximum pressure under hypothetical accident conditions. All information presented in this table should be consistent with the information used within other sections of the application.

3.2 Material Properties and Component Specifications

3.2.1 Material Properties

This section should specify the appropriate thermal properties for materials that affect heat transfer both within the package and from the package to the environment. Liquids or gases within the package and gases external to the package for hypothetical accident conditions should be included. For packages using anisotropic materials, the directional properties of these materials should be provided. The thermal absorptivities and emissivities should be appropriate for the package surface conditions and each thermal condition. When reporting a property as a single value, the evaluation should show that this value bounds the equivalent temperature-dependent property. In addition, this section should include references for the data provided.

These properties include the following:

1.   Thermal conductivity;

2.   Specific heat;

3.   Density;

4.   Thermal radiation emissivity of the package surfaces;

5.   Coefficient of thermal expansion; and

6.   Modulus of elasticity.

3.2.2 Component Specifications

This section should include the technical specifications of components that are important to the thermal performance of the package as illustrated by the following examples:

1.   In the case of valves or seals, the operating pressure range and temperature limits;

2.   The properties of fabricated insulation and coatings, including a summary of test data that support their performance specifications;

3.   The maximum allowable service temperatures or pressures for each package component, including pressure relief valves and fusible plugs; and

4.   The minimum allowable service temperature of all components, which should be less than or equal to $-40°$ C ($-40°$ F).

3.3 General Considerations

The thermal evaluation of the package can be performed by either analysis or test, or a combination of both.

3.3.1 Evaluation by Analysis

For computer analyses, the method used should be properly referenced or developed, and the computer program should be shown to be well benchmarked for thermal analyses, applicable to the evaluation, and sufficiently described to permit review and independent verification. The assumptions used in modeling the heat sources and heat transfer paths should be clearly stated and justified.

The thermal analysis should assume that the heat transfer medium is air, and the effects of air on the contents and packaging components (e.g., oxidation of depleted uranium shielding) should be considered. For packages using other fill gas (e.g., argon, helium), the analysis should assume air and may also include an analysis with the actual fill gas to show the impact on the thermal performance of the package.

The analysis should include the following:

1.   Thermal properties for the package materials;

2.   Calculations for conductive, convective, and radiative heat transfer among package components and from the surfaces of the package to the environment;

3.   A description of the changes in package geometry and material properties resulting from structural and thermal tests under normal conditions of transport and hypothetical conditions of transport;

4. The heat produced by the combustion of package components, if applicable;

5. A description of the temperature and thermal boundary conditions for normal conditions of transport and hypothetical conditions of transport; and

6. A demonstration that the time interval used for the evaluation of the temperature following the thermal test is adequate to ensure that the components reach their maximum temperature and that steady-state temperatures have been reached.

For the 30-minute fire test, the majority of the heat input to the package will be through radiation. For convective heat transfer, a convective heat transfer coefficient appropriate for the conditions that would exist if the package were exposed to the fire should be used. Flame velocities in an open pool fire may be used in determining the appropriate convective heat transfer coefficient. Any correlation used in the analysis should be properly explained and justified. For the post-fire cooldown, natural convection should be assumed.

Any assumptions for contact resistance at material interfaces, energy transport across gaps, enclosures, and other factors should be provided and justified.

For SNF packages, an analysis should be considered to evaluate the potential impact of the fission gas to the cask component temperature limits and the cask internal pressurization if any of the following conditions are met:

1. The cask component temperatures are within 5 percent of their limiting values under accident conditions;

2. The maximum normal operation pressure is within 10 percent of its design-basis pressure; or

3. Any other special conditions exist.

3.3.2 Evaluation by Test

The evaluation should include a detailed description of the design of the test package and test facility. This description should demonstrate that the test package was fabricated under a proper quality assurance program. In addition, the evaluation should include a description of how the test facility operates and provide details on how the results were evaluated.

The application should include the following:

1. Demonstration that the test facility (pool fire or furnace facility) and the test procedure can meet the range of thermal conditions such as fire heat fluxes or temperature;

2. A description of the performance of the test package, including the simulated content and any attached test instrumentation and mounting hardware;

3. Demonstration that the temperature-sensing instrumentation has been located to measure the maximum temperature of the package components and can adequately characterize the heat transfer pathways; and

4.   Demonstration that the package instrumentation, such as temperature- or pressure-sensing devices, has been mounted in locations that minimize its effects on local test package temperatures.

Some conditions, such as ambient temperature, decay heat of the contents, or package emissivity or absorptivity, may not be exactly represented in a thermal test. The thermal evaluation should include appropriate corrections or evaluations to account for these differences. For example, the thermal evaluation should include a temperature correction if the ambient temperature at the onset of the fire test was lower than 38° C (100° F).

3.4 Thermal Evaluation under Normal Conditions of Transport

This section should describe the thermal evaluation of system and subsystem operation under normal conditions of transport. The temperature ranges bounded by the minimum and maximum ambient temperatures and minimum and maximum decay heat loads should be considered. The results should be compared with allowable limits of temperature and pressure for the package components. The information should be presented in summary tables along with statements and appropriate comments. Information that is to be used in other sections of the review should be identified. The margins of safety for package temperatures, pressures, and thermal stresses, including the effects of uncertainties in thermal properties, test conditions and diagnostics, and analytical methods, should be addressed. The analysis or test results should be shown to be reliable and repeatable.

In addressing the sections below, the following general information should be considered and included as appropriate:

1.   Assumptions that are used in the analysis should be clearly described and justified.

2.   Models and modeling details should be clearly described.

3.   For thermal evaluation by test, the test method, procedures, equipment, and facilities that were used should be described in detail.

4.   If the specimen tested is not identical in all respects to the package described in the application, the differences should be described and justification given that these differences would not affect the test results.

5.   Temperature data should be reported at gaskets, valves, and other containment boundaries, particularly for temperature-sensitive materials, as well as for the overall package; and

6.   Both interior and exterior temperatures should be included.

The damage caused by the tests and the results of any measurements that were made should be reported in detail, including photographs of the testing and the test specimen.

3.4.1 Heat and Cold

This section should demonstrate that the tests for normal conditions of transport do not result in a significant reduction in packaging effectiveness. The following items should be considered and addressed:

1. Degradation of the heat-transfer capability of the packaging, such as creation of new gaps between components;

2. Changes in material conditions or properties (e.g., expansion, contraction, gas generation, and thermal stresses) that affect the structural performance;

3. Changes in the packaging that affect containment, shielding, or criticality, such as thermal decomposition or melting of materials; and

4. Ability of the packaging to withstand the tests under hypothetical accident conditions.

The component temperatures and pressures should be compared to their allowable values. This section should explicitly show that the maximum temperature of the accessible package surface is less than 50° C (122° F) for nonexclusive-use shipment or 85° C (185° F) for exclusive-use shipment in accordance with the requirements specified in 10 CFR 71.43(g) or Paragraphs 652 and 662 of TS-R-1 which are incorporated in Subsection 1(1) of the PTNS Regulations by reference to Paragraph 650 of TS-R-1.

3.4.2 Temperatures Resulting in Maximum Thermal Stresses

The evaluation of thermal stresses caused by constrained interfaces among package components resulting from temperature gradients and differential thermal expansions should be addressed in Sections 2.6.1.2 and 2.6.1.3 of the application.

3.4.3 Maximum Normal Operating Pressure

This section should report the maximum normal operating pressure consistent with the other sections of the safety analysis report and show how it was calculated, assuming the package has been subjected to the heat condition for 1 year as specified in 10 CFR 71.4 and 10 CFR 71.33(b)(5) or Paragraph 660 of TS-R-1 which is incorporated in Subsection 1(1) of the PTNS Regulations by reference to Paragraph 650 of TS-R-1. The calculation should consider possible sources of gases, including the following:

1. Gases initially present in the package;

2. Saturated vapor, including water vapor from the contents or packaging;

3. Helium from the radioactive decay of the contents;

4. Hydrogen or other gases resulting from thermal or radiation-induced decomposition of materials such as water or plastics; and

5. Fuel rod failure.

For SNF packages, Table 4-1 provides guidance on release of fill gas and fission product gas for pressurized and boiling light-water reactor fuel.

This section should also address the requirement in 10 CFR 71.4 or Paragraph 661 of TS-R-1, which is incorporated in Subsection 1(1) of the PTNS Regulations by reference to Paragraph 650 of TS-R-1, with respect to Type B(U) packages (i.e., the maximum normal operating pressure must not be greater than 700 kPa (100 lbf/in.^{2}) gauge pressure).

In addition, this section should demonstrate that hydrogen and other flammable gases comprise less than 5 percent by volume of the total gas inventory within any confined volume and will not result in a flammable mixture within any confined volume of the package as specified in 10 CFR 71.43(d) or Paragraph 642 of TS-R-1 which is incorporated in Subsection 1(1) of the PTNS Regulations by reference to Paragraph 650 of TS-R-1.

3.5 Thermal Evaluation under Hypothetical Accident Conditions

This section should describe the thermal evaluation of the package under hypothetical accident conditions. The hypothetical accident conditions defined in 10 CFR 71.73 or Paragraphs 726–729 of TS-R-1 should be applied sequentially to meet 10 CFR 71.73 or, in the most damaging sequence, to meet TS-R-1 (see Section 2.7 above). For the accident condition thermal evaluation, the general comments in Section 3.3 above should be considered and addressed as appropriate.

3.5.1 Initial Conditions

The thermal evaluation should consider the effects of the drop, crush (if applicable), and puncture tests on the package. This section should identify initial conditions, including the following, and justify that they are most unfavorable:

1.	An ambient temperature between −40° C (−40° F) and +38° C (+100° F) with no insolation. This range is specified in the Canadian and IAEA regulations. The U.S. regulations specify a minimum ambient temperature of −29° C (−20° F) for the initial condition. Therefore, an ambient temperature range between −40° C (−40° F) and +38° C (+100° F) should be considered.

2. An internal pressure of the package equal to the maximum normal operating pressure unless a lower internal pressure, consistent with the ambient temperature, is less favorable.

3. Contents at its maximum decay heat unless a lower heat, consistent with the temperature and pressure, is less favorable.

3.5.2 Fire Test Conditions

This section should provide a detailed description of the analysis or tests used to evaluate the package under the fire test conditions. The evaluation should address the requirements in 10 CFR 71.73(c)(4) or Paragraph 728 of TS-R-1 which is incorporated in Subsection 1(4) of the PTNS Regulations by reference to Paragraph 716 of TS-R-1.

The package should be subjected to full insolation, and it should be ensured that the evaluation is continued until the post-fire, steady-state condition is achieved and that no artificial cooling is applied to the package. In addition, all combustion should be allowed to proceed until it terminates naturally.

When a thermal test is performed to evaluate the package performance, the description of the test should include the following:

1. Fire dimensions;

41

2.   Package orientation and support methods;

3.   Test temperatures and duration;

4.   Heat source;

5.   Initial ambient temperature;

6.   Period following the thermal test; and

7.   Adequate availability of oxygen supply.

For a pool fire, the fire width should extend horizontally between 1 m (40 in.) and 4 m (10 ft) beyond any external surface on the package. In addition, the package should be positioned 1 m (40 in.) above the surface of the fuel source.

The analysis should include the thermal performance of the test package, including simulated package contents and any attached test instrumentation and mounting hardware, including the location of temperature-sensing instrumentation used to measure the appropriate maximum package component temperatures and characterize the significant heat transfer pathways. These should be mounted at locations that minimize their effects on local test package temperatures. Any possible perturbations caused by the presence of these sensors should be appropriately considered.

Any physical change in the package condition resulting from the fire test, such as change in material properties, combustion or melting of package components, and increase in internal temperature and pressure during the fire and post-fire period, should be adequately evaluated and justified.

3.5.3 Maximum Temperatures and Pressure

This section should report the peak temperatures of package components as a function of time, both during and after the fire, as well as the maximum temperatures from the post-fire, steady-state condition. This section should include those temperatures at locations in the package that are significant to the safety analysis and review. In particular, the temperatures for items such as contents, gaskets, valves, and shielding should be reported. These temperatures should not exceed their maximum allowable values; melting of lead shielding is not permitted. The calculations of temperatures should trace the temperature-time history up to and past the time at which maximum temperatures are achieved and begin to fall.

The evaluation of the maximum pressure in the package should be based on the maximum normal operating pressure and should consider fire-induced increases in package temperatures, thermal combustion or decomposition processes, fuel rod failure, phase changes, and other factors.

This section should provide a general description of package performance and should compare the results of the thermal test with allowable limits of temperature, pressure, and other characteristics for the package components. Damage to the package either from interpretation of the analysis or from test observation should be considered or described. The assessment should include possible structural damage, breach of containment, and loss of shielding.

42

3.5.4 Temperatures Resulting in Maximum Thermal Stresses

This section should present results of thermal analyses used in the structural evaluation for calculating the most severe thermal stress conditions that result during the fire test and subsequent cooldown. The temperatures corresponding to the maximum thermal stresses should be reported.

3.5.5 Fuel/Cladding Temperatures for Spent Nuclear Fuel

For SNF packages, the maximum allowable fuel/cladding temperature should be identified and justified. The justification should consider the fuel and clad materials, irradiation conditions (e.g., the absorbed dose, neutron spectrum, and fuel burnup), and the shipping environment, including the fill gas. In general, the cladding temperature for commercial light-water-reactor spent fuel should be kept below 400° C (752° F) under normal conditions of transport and below 570° C (1,058° F) under accident conditions. Other necessary considerations include the hydride reorientation effects on mechanical properties, the elapsed time from the removal of the SNF from the core to its placement into the transportation packaging, its duration in the packaging, and its post-transport disposition. Examples of temperature limits include the following:

1.   The temperature limit for metal fuel, which should be less than the lowest melting point eutectic of the fuel; and

2.   The temperature limit on the irradiated clad in an inert gas environment as appropriate.

3.5.6 Accident Conditions for Fissile Material Packages for Air Transport

If applicable, applicants should address the expanded fire test conditions specified in 10 CFR 71.55(f)(1)(iv) or Paragraph 736 of TS-R-1 which is incorporated in Subsection 1(4) of the PTNS Regulations by reference to Paragraph 716 of TS-R-1.

3.6 Appendix

The appendix should include a list of references, applicable pages from referenced documents, justification of assumptions or analytical procedures, test results, photographs, computer program descriptions and examples of input and output files, specifications of O-rings and other components, detailed materials test data, and other supplemental information.

If the package has been subjected to a thermal test, the appendix should include a description of the test facility with respect to the following:

1.   Type of facility (e.g., furnace, pool fire);

2.   Method of heating the package (e.g., gas burners, electrical heaters);

3.   Volume and emissivity of the furnace interior;

4.   Method of simulating decay heat, if applicable;

5. Types, locations, and measurement uncertainties of all sensors used to measure the fire heat fluxes affecting critical components, such as seals, valves, pressure, and structural components, and fire temperatures;

6. The post-fire environment for a period adequate to attain the post-fire, steady-state condition; and

7. Methods for both maintaining and measuring an adequate supply and circulation of oxygen for initiating and naturally terminating the combustion of any burnable package component throughout the fire and post-fire periods.

The appendix should also include a complete description of the tests performed. This description should include the following:

1. Test procedure;

2. Test package description;

3. Test initial and boundary conditions;

4. Test chronologies (planned and actual);

5. Photographs of the package components, including any structural or thermal damage, before and after the tests;

6. Test measurements, including, at a minimum, documentation of test package physical changes and temperature and heat flux histories;

7. Corrected tests results; and

8. Method used to obtain the corrected results.

4. Containment

This section of the application should identify the package containment system and describe how the package complies with the containment requirements of 10 CFR 71.43(f) and 10 CFR 71.51, "Additional Requirements for Type B Packages," or Paragraphs 646, 656, 658, and 659 of TS-R-1 which are incorporated in Subsection 1(1) of the PTNS Regulations by reference to Paragraph 650 of TS-R-1.

The section should address the structural and thermal effects on the packaging and its content under normal and hypothetical accident conditions and their effects on the containment system of the package. Any operational, fabrication, and maintenance requirements with respect to containment for the package should be included in the application in Sections 7, "Package Operations," and 8, "Acceptance Tests and Maintenance Program."

4.1 Description of the Containment System

This section should define and describe the containment system. The containment boundary of the package should be explicitly identified, including the containment vessel, welds, drain or fill ports, valves, seals, test ports, pressure relief devices, lids, cover plates, and other closure devices. If multiple seals are used for a single closure, this section should identify the seal defined as the containment system seal. Detailed drawings of the containment system should be included.

Packaging design features important for containment include the following:

1.   Materials of construction of the containment system;

2.   Welds;

3.   Applicable codes and standards (e.g., ASME Code specifications for the vessel);

4.   Bolt torque required to maintain positive closure;

5.   Maximum and minimum allowable temperatures of components, including seals; and

6.   Maximum and minimum temperatures of components under the tests for normal conditions and hypothetical accident conditions of transport.

All containment boundary penetrations and their method of closure should be adequately described. Performance specifications for components such as valves, O-rings, and pressure-relief devices should be identified and documented; no device may allow continuous venting.

The containment evaluation should show that compliance with the containment requirements does not rely on any filter or mechanical cooling system, as specified in 10 CFR 71.51(c) or Paragraph 659 of TS-R-1 which is incorporated in Subsection 1(1) of the PTNS Regulations by reference to Paragraph 650 of TS-R-1.

If the design includes valves or similar devices, the application should demonstrate that these are protected against unauthorized operation and, except for a pressure-relief valve, have an enclosure to retain any leakage.

Demonstration that no galvanic, chemical, or other reactions will occur between the seal and the packaging or its contents, and that the seal will not degrade from irradiation, should be addressed. If penetrations are closed with two seals, specification of which of the two seals is defined as the containment boundary should be addressed.

Specifications of the seal grooves and the type and size of seals should be provided. The temperature of containment boundary seals should be shown to remain within the specified allowable limits under both normal conditions and hypothetical accident conditions of transport.

Demonstration of how the containment system is securely closed with a positive fastening device that cannot be opened unintentionally or by pressure that may arise within the package and a description of the features that ensure that continuous venting is precluded should be included.

Scale-model testing is not a reliable or acceptable method for qualifying the leakage rate of a full-scale package. If compliance is demonstrated by analysis, the structural evaluation should show that the containment boundary, seal region, and closure bolts do not undergo any inelastic deformation and that the materials of the containment system (e.g., seals) do not exceed their maximum allowable temperature limits.

For SNF packages, the material used for the containment system and the design, fabrication, examination, testing, inspection, and certification should be in accordance with Section III, Division 3, of the ASME Boiler and Pressure Vessel Code (Ref. 9). This includes an agreement with an authorized inspection agency to provide inspection and audit services for the design owners, packaging owners, and Class W certificate holders. Justification for the use of other codes should be provided in the application. Also, the codes, standards, and criteria for the inner containment system should generally be the same as those of the outer containment system. Justification for differences should be presented in the application.

4.1.1 Special Requirements for Damaged Spent Nuclear Fuel

For damaged SNF, the determination of the fuel condition should be based, at a minimum, on review of fuel records. Damaged fuel may consist of fuel assemblies with either cladding or structural defects. Fuel with damaged cladding should be contained to facilitate handling and to confine gross fuel particles to a known subcritical configuration under normal and hypothetical accident conditions. Use of a canister for the fuel may be an option for consideration. The application should include justification for the material specifications and the design/fabrication criteria for the canister. These specifications and criteria should be the same as those for containment or criticality support structures.

4.2 Containment under Normal Conditions of Transport

This section should include the evaluation of the containment system under normal conditions of transport, using the methods in American National Standards Institute (ANSI) N14.5, "American National Standard for Radioactive Materials—Leakage Tests on Packages for Shipment" (Ref. 15), or International Standardization Organization (ISO) 12807, "Safe Transport of Radioactive Materials—Leakage Testing on Packages" issued in 1996 (Ref. 16). This section should demonstrate that the package meets the containment requirements of 10 CFR 71.51(a)(1) or Paragraph 656(a) of TS-R-1 which is incorporated in Subsection 1(1) of the PTNS Regulations by reference to Paragraph 650 of TS-R-1 under normal conditions of transport. The evaluation should be performed for the most limiting chemical and physical forms of the contents. Significant daughter products should be included. The constituents of the releasable source term, including radioactive gases, liquids, and powder aerosols, should be identified. If less than 100 percent of the contents are considered releasable, a justification for the lower fraction should be included. The containment evaluation should not rely on the blockage of a leakage path by particulate contents to meet the containment criteria in the regulations. Any seal demonstrating a leakage rate of 1×10^{-7} reference-cubic centimeters per second (cm^3/s), as defined in ANSI N14.5, may be considered to be leaktight.

The evaluation under normal condition of transport should include the following:

1. The maximum internal pressures, including any gases generated in the package during a period of 1 year;

2. The structural performance of the containment system, including seals, closure bolts, and penetrations; and

3. The leak testing of the containment system.

Combustible gases should not exceed 5 percent (by volume) of the free gas volume in any confined region of the package. No credit should be taken for getters, catalysts, or other recombination devices.

For Type A fissile material packages, the evaluation should show that there is no loss or dispersal of radioactive material under normal conditions of transport, as specified in 10 CFR 71.43(f) or Paragraph 646 of TS-R-1 which is incorporated in Subsection 1(1) of the PTNS Regulations by reference to Paragraph 650 of TS-R-1. For Type B packages, the evaluation should show that there is no release under normal conditions of transport to the required sensitivity. In both cases, there should be no significant increase in external radiation levels.

For SNF packages, the releasable source term is composed of crud (surface contamination) on the outside of the fuel rod cladding that can become aerosolized and the fuel fines, volatiles, and gases that are released from a fuel rod in the event of a cladding breach. Bounding values for the effective surface activity density in becquerels per square centimeter (Bq/cm^2) (Ci/cm^2) of the crud on fuel rod cladding are based on experimental determinations. A computer code, such as ORIGEN-S, can be used to identify the radionuclides present for a given percent fuel enrichment, burnup, and cooldown time. Using the individual A_2 values for the crud, fines, gases, and volatiles individually, the effective A_2 of the releasable source-term mixture can be determined by using the relative release fraction for each contributor and the methods from ANSI N14.5 (Ref. 15) or ISO 12807 (Ref. 16). The release fractions and effective specific activities for the various releasable source-term contributors for SNF with an initial enrichment of 3.2 percent, a burnup of 33 gigawatt-days per metric ton initial heavy metal (GWD/MTIHM), and a cooldown time of 5 years are given in Table 4-1. The release fractions presented in Table 4-1 have been developed from reasoned argument and experimental data (NUREG/CR-6487, "Containment Analysis for Type B Packages Used to Transport Various Contents," issued November 1996 (Ref. 17)). These values may be considered as default values for light-water reactor SNF with a burnup below 45 GWD/MTIHM. The release fractions and specific activities should be justified in the application as appropriate.

Table 4-1 Release Fractions and Specific Activities for the Contributors to the Releasable Source Term for Packages Designed to Transport Irradiated Fuel from Commercial Pressurized- and Boiling-Water Reactors for Burnups below 45 GWD/MTIHM

Variable	PWR		BWR	
	Normal Conditions of Transport	Hypothetical Accident Conditions	Normal Conditions of Transport	Hypothetical Accident Conditions
Fraction of crud that spalls off of rods, f_C	0.15	1.0	0.15	1.0
Crud surface activity, S_C (Ci/cm^2)*	140×10^{-6}	140×10^{-6}	1254×10^{-6}	1254×10^{-6}
Mass fraction of fuel that is released as fines because of a cladding breach, f_F	3×10^{-5}	3×10^{-5}	3×10^{-5}	3×10^{-5}
Specific activity of fuel rods, A_R (Ci/g)	0.60	0.60	0.51	0.51
Fraction of rods that develop cladding breaches, f_B	0.03	1.0	0.03	1.0
Fraction of gases that are released because of a cladding breach, f_G	0.3	0.3	0.3	0.3
Specific activity of gas in fuel rod, A_G (Ci/g)	7.32×10^{-3}	7.32×10^{-3}	6.28×10^{-3}	6.28×10^{-3}
Specific activity of volatiles in a fuel rod, A_V (Ci/g)	0.1375	0.1375	0.1794	0.1794
Fraction of volatiles that are released due to a cladding breach, f_V	2×10^{-4}	2×10^{-4}	2×10^{-4}	2×10^{-4}

* Values for the crud activity are for the time of reactor discharge and should be corrected for radioactive decay.

The maximum permissible release rate and the maximum permissible leakage rate should be based on the mass density, effective specific activity, and effective A_2 of the releasable source term and should be calculated in accordance with the methods specified in ANSI N14.5 (Ref. 15) or ISO 12807 (Ref. 16).

4.3 Containment under Hypothetical Accident Conditions

This section should include the evaluation of the containment system under hypothetical accident conditions, considering factors given in Section 4.2 above. This section should demonstrate that the package meets the containment requirements of 10 CFR 71.51(a)(2) or Paragraph 656(b) of TS-R-1 which is incorporated in Subsection 1(1) of the PTNS Regulations by reference to Paragraph 650 of TS-R-1 under hypothetical accident conditions. In particular, the structural performance of the containment system should be addressed, including seals, closure bolts, and penetrations, as well as leakage testing of the containment system. The evaluation should consider differences relevant to the accident conditions (e.g., pressurization of the containment system under fire test conditions, a possible increase in the releasable source term, and possible changes in containment system performance from package damage).

4.4 Leakage Rate Tests for Type B Packages

This section should describe leakage tests that are used to show that the package meets the containment requirements of 10 CFR 71.51 or Paragraph 656 of TS-R-1 which is incorporated in Subsection 1(1) of the PTNS Regulations by reference to Paragraph 650 of TS-R-1. These may include the following:

1.   Fabrication leakage rate test;

2.   Maintenance leakage rate test;

3.   Periodic leakage rate test; and

4.   Pre-shipment leakage rate test.

Fabrication, maintenance, and periodic leakage rate tests should be included in Section 8, "Acceptance Tests and Maintenance Program," of the application. The pre-shipment leakage rate test for assembly verification should be included in the operating procedures, as detailed in Section 7, "Package Operations," of the application.

Sample analyses for determining containment criteria for Type B packages are provided in NUREG/CR-6487 (Ref. 17). If these analyses are used, the applicant should demonstrate that the assumptions used in NUREG/CR-6487 are applicable to the package under consideration.

Methods for leak testing of all containment seals and penetrations, including drain and vent ports, should be described. If fill, drain, or test ports utilize quick-disconnect valves, demonstration that these do not preclude leak testing of the containment seals should be provided. The maximum allowable leakage rate and the minimum test sensitivity should be specified for each type of test (e.g., fabrication, maintenance, periodic, and pre-shipment tests).

A method to determine the maximum permissible volumetric leakage rates based on the allowed regulatory release rates under both normal conditions of transport and hypothetical accident conditions can be found in ANSI N14.5 (Ref. 15). The smaller of these air-leakage rates is defined as the reference air-leakage rate. ISO 12807 (Ref. 16) also presents a way of calculating the release rate.

4.5 Appendix

The appendix should include a list of references, applicable pages from referenced documents, supporting information and analysis, test results, and other appropriate supplemental information.

5. Shielding Evaluation

This section of the application should identify, describe, discuss, and analyze the principal radiation shielding design of the packaging, components, and systems that are important to safety. This section should address the regulatory requirements of 10 CFR 71.47, "External Radiation Standards for All Packages," and 10 CFR 71.51(a)(1) and (2) or Paragraphs 526, 530, 531, 532 of TS-R-1 as referenced in Subsection 16(4) of the PTNS Regulations, Paragraph 572 of TS-R-1 as referenced in Subsection 15(5) of the PTNS Regulations, and Paragraphs 645, 646(b), and 656(b)(ii)(i) of TS-R-1 which are incorporated in Subsection 1(1) of the PTNS Regulations by reference to Paragraph 650 of TS-R-1.

5.1 Description of Shielding Design

5.1.1 Design Features

This section should describe the radiation shielding design features of the package. Design features important to shielding include the following:

1. Dimensions, tolerances, and densities of material for neutron or gamma shielding, including those packaging components considered in the shielding evaluation;

2. Mass density, atomic density, or area density of materials used as neutron absorbers;

3. Methods used to determine the uniformity of the absorbers along with support references to the data;

4. Structural components that maintain the contents in a fixed position within the package; and

5. Dimensions of the transport vehicle that are considered in the shielding evaluation.

The text, tables, and figures describing the shielding design features should be consistent with the engineering drawings and the models used in the shielding evaluation.

5.1.2 Summary Table of Maximum Radiation Levels

This section should present the maximum dose rates for both normal conditions of transport and hypothetical accident conditions at the appropriate locations for nonexclusive- and exclusive-use shipments, as applicable. Tables 5-1 and 5-2 present an appropriate format for providing the external radiation information specific to the packaging with its contents for shipments in non-exclusive use vehicles, and in exclusive use vehicles, respectively.

For SNF packages, the spent fuel specification (e.g., burnup, enrichment, and cooling time) at which the individual radiation levels apply should be given in the table since the gamma or neutron contributions could be greatest at different fuel specifications.

Table 5-1 Summary Table of External Radiation Level (Nonexclusive Use)

	Package Surface mSv/h (mrem/h)			1 Meter from Package Surface mSv/h (mrem/h)		
Normal Conditions of Transport	Top	Side	Bottom	Top	Side	Bottom
Gamma						
Neutron						
Total						
10 CFR 71.47(a) or Paragraphs 530 and 531 of TS-R-1 limit	2 (200)	2 (200)	2 (200)	0.1 (10)*	0.1 (10)*	0.1 (10)*
Hypothetical Accident Conditions						
Gamma						
Neutron						
Total						
10 CFR 71.51(a)(2) or Paragraph 656(b)(ii)(I) of TS-R-1 limit				10 (1000)	10 (1000)	10 (1000)

* Transport index may not exceed 10.

Table 5-2 Summary Table of External Radiation Level (Exclusive Use)*

	Package (or Freight Container) Surface mSv/h (mrem/h)			2 Meters from Outer Vehicle Surface mSv/h (mrem/h)		
Normal Conditions of Transport	Top	Side	Bottom	Top	Side	Bottom
Gamma						
Neutron						
Total						
10 CFR 71.47(b) or Paragraph 572 of TS-R-1 limit	10 (1000)**	10 (1000)**	10 (1000)**	0.1 (10)	0.1 (10)	0.1 (10)
	Vehicle Surface mSv/h (mrem/h)			**Occupied Position mSv/h (mrem/h)**		
Normal Conditions of Transport	Top	Side	Underside			
Gamma						
Neutron						
Total						
10 CFR 71.47(b) or Paragraph 572 of TS-R-1 limit	2 (200)	2 (200)	2 (200)	0.02 (2)		

Hypothetical Accident Conditions	1 Meter from Package Surface mSv/h (mrem/h)		
Gamma			
Neutron			
Total			
10 CFR 71.51(a)(2) or Paragraph 656(b)(ii)(I) of TS-R-1 limit	10 (1000)	10 (1000)	10 (1000)

* For packages transported by roadway, railway, and sea
** For packages in closed vehicles; otherwise, 2 (200)

5.2 Source Specification

This section should describe the contents as well as the gamma and neutron source terms used in the shielding analysis. Any increase in source terms over time should be considered. For those packages designed for multiple types of contents, the contents producing the highest external dose rate at each location should be clearly identified and evaluated. For packages designed for spent fuel transport, this section should also state the fuel type, fuel burnup, cooling time, and initial enrichment. For spent fuel package shielding evaluations, the neutron source term increases considerably with decreasing initial enrichment and constant burnup. Consequently, in identifying the bounding source term, the minimum initial enrichment should be specified. Note that the appropriate cross-section for the corresponding spent fuel burnup should be used.

5.2.1 Gamma Source

This section should specify the quantity of radioactive material included as contents and tabulate the gamma decay source strength (megaelectronvolts per second (MeV/s) and photons/s) as a function of photon energy. A detailed description of the method used to determine the gamma source strength and distribution should be provided.

For non-SNF contents, the maximum gamma source strength and spectra should be calculated by an appropriate method (e.g., standard computer codes or hand calculations). The source contribution from radioactive daughter products should be included if it produces higher dose rates than the contents without decay. If the radioactive nuclides and gamma spectra are calculated with a computer code, the key parameters should be described in the application or listed in the input file. The production of secondary gammas (e.g., from (n,γ) reactions in shielding material) should be either calculated as part of the shielding evaluation (see Section 5.4 of this document) or otherwise appropriately included in the source term.

The results of the source-term determination should be presented as a listing of gammas per second, or MeV/s, as a function of energy. The activity (or mass) of each nuclide that contributes significantly to the source term should also be provided as supporting information.

For SNF contents, the gamma source terms should be specified as a function of energy for both the SNF and activated hardware. If the energy group structure of the source-term calculation differs from that of the cross-section set of the shielding calculation, the applicant may need to regroup the photons. In general, only gammas from approximately 0.8 to 2.5 MeV will contribute significantly to the external radiation levels, so regrouping outside of this range is of little consequence. A consistent source-term unit (e.g., per assembly, per total number of assemblies, or per metric ton) should be used in the shielding calculation.

Determining the source terms for fuel assembly hardware is generally not as straightforward as that for SNF. The activation of the hardware depends on the impurities (e.g., cobalt-59) initially present and on the spatial and energy variation of the neutron flux during burnup. If the package is intended to transport other hardware, such as control assemblies or shrouds, the source terms from these components should be included.

Depending on the packaging design, neutron interactions could result in the production of energetic gammas near the packaging surface. If the shielding analysis code does not treat this source code, other appropriate means should be used for its determination.

5.2.2 Neutron Source

This section should specify the quantity of radioactive material included as contents and tabulate the neutron source strength (neutrons per second) as a function of energy. A detailed description of the method used to determine the neutron source strength and distribution should be provided.

The method should consider, as appropriate, neutrons from both spontaneous fission and from (α,n) reactions. Depending on the methods used to calculate these source terms, the applicant might determine the energy group structure independently. This is often accomplished by selecting the nuclide with the predominant contribution to spontaneous fission (e.g., curium-244) and using that spectrum for all neutrons since the contribution from (α,n) reaction is generally small. If either of these source contributions is assumed to be negligible, an appropriate justification should be provided.

The production of neutrons from subcritical multiplication should be either calculated as part of the shielding evaluation (see Section 5.4 of this document) or otherwise conservatively included and justified in the source term.

The results of the source-term calculation, if applicable, should be presented as a listing of neutrons per second as a function of energy. The contribution from spontaneous fission and (α,n) should be separately identified along with the actinides or light nuclei significant for these processes. For the spontaneous fission contribution, a listing of the significant nuclides should also be presented.

5.3 Shielding Model

5.3.1 Configuration of Source and Shielding

This section should provide a detailed description of the model used in the shielding evaluation. The effects of the tests on the packaging and its contents under normal conditions of transport and hypothetical accident conditions should be evaluated. The models used in the shielding calculation should be consistent with these effects.

This section should include sketches (to scale) and dimensions of the radial and axial shielding materials. The dimensions of the transport vehicle and the package location for exclusive-use shipments should be included for the purpose of determining the radiation level at 2 m (80 in.) from the vehicle and the normally driver-occupied locations. The analysis is based on the radiation limits in 10 CFR 71.47(b) or Paragraph 572 of TS-R-1 as referenced in Subsection 15(5) of the PTNS Regulations.

The dose point locations in the shielding model, including all locations prescribed in 10 CFR 71.47(a) or 10 CFR 71.47(b), plus 10 CFR 71.51(a)(2), or Paragraphs 530 and 531 of TS-R-1 as referenced in Subsection 16(4) of the PTNS Regulations, or Paragraph 572 of TS-R-1 as referenced in Subsection 15(5) of the PTNS Regulations, and Paragraph 656(b)(ii)(i) of TS-R-1 which is incorporated in Subsection 1(1) of the PTNS Regulations by reference to Paragraph 650 of TS-R-1, should be identified. These points should be chosen to identify the locations of the maximum radiation levels. Radiation peaking often occurs near the edges of external neutron shield and impact limiters for SNF packages. Voids, streaming paths, and irregular geometries in the model should be included or otherwise treated in a conservative manner.

If contents can be positioned at varying locations or with varying densities, the location and physical properties of the contents used in the evaluation should be those resulting in the maximum external radiation levels. For example, the source configuration that maximizes radiation level on the side of the package might not be the same source configuration that maximizes the radiation level on the top or bottom. Any changes in configuration (e.g., displacement of source or shielding, reduction in shielding) resulting under normal conditions of transport or hypothetical accident conditions should be included as appropriate.

For SNF packages, the source-term locations for both SNF and the structural support regions of the fuel assemblies should be modeled properly. Generally, at least three source regions (fuel and top/bottom assembly hardware) are necessary. Within the SNF region, the fuel materials may generally be homogenized to facilitate shielding calculations. In some cases, the basket material may be homogenized also. However, homogenization may not be appropriate in some cases when it distorts the neutron multiplication rate or when radiation streaming can occur between the basket components. In addition, the assumed source configuration should bound damaged conditions of the spent fuel assemblies if damaged fuels are to be loaded in the packaging.

Because of the burnup profile for SNF, a uniform source distribution is generally conservative for the top and bottom dose points but not for the axial center unless the source strength is appropriately adjusted. If peaking appears to be significant, it should be treated appropriately. The assembly structural support regions (e.g., top/bottom end pieces and plenum) should be correctly positioned relative to the SNF. These support regions may be individually homogenized.

5.3.2 Material Properties

This section should describe the material properties (e.g., mass densities and atom densities) in the shielding models of the packaging and contents. Changes resulting under normal conditions of transport or hypothetical accident conditions should be included as appropriate. The sources of data for uncommon materials should be cited. The uncommon materials should be properly controlled to achieve their design densities. Specific information on control measures should be included in Section 8, "Acceptance Tests and Maintenance Program" of the application.

Shielding properties of the materials should not degrade during the service life of the packaging (e.g., degradation of foam or dehydration of hydrogenous materials). Controls should be in place to ensure the long-term effectiveness of the shielding as appropriate. Temperature-sensitive shielding materials should not be subject to temperatures at or above their design limitations during either normal or accident conditions. The applicant should properly examine the potential for shielding materials that may experience changes in material densities at temperature extremes. For example, elevated temperatures may reduce hydrogen content through loss of bound or free water in hydrogenous shielding materials. In addition, temperatures that may result in changing the form of shielding materials, such as melting of lead shielding, is not acceptable.

If the shielding model considers a homogenous source region (rather than a detailed heterogeneous model of the contents), such an approach should be justified, and it should be demonstrated that the homogenized mass densities are correct for normal conditions of transport and hypothetical accident conditions. Atom densities should also be confirmed if used as input to shielding calculations.

5.4 Shielding Evaluation

5.4.1 Methods

This section should provide a general description of the basic method used to determine the gamma and neutron dose rates at the selected points outside the package for both normal and accident conditions of transport. This should include a description of the spatial source distribution and any computer program used, with its referenced documentation. This section should also include a detailed description of the basic input parameters as well as the bases for selecting the program, attenuation and removal cross-sections, and buildup factors.

The computer codes may use Monte Carlo transport, deterministic transport, or point-kernel techniques. The latter is generally appropriate only for gammas. For computer codes not well established in the public domain, the application should describe the solution method, benchmark results, validation procedure, and QA practices.

The dimensions of the modeling/code (one-dimensional, two-dimensional, or three-dimensional) should be commensurate with the complexity of the package and the content. Generally, for an SNF package, a two-dimensional or three-dimensional calculation is necessary. One-dimensional codes provide little information about off-axis locations and streaming paths. Even for radiation levels at the end of the package, one-dimensional codes require a buckling correction that must be justified; merely using the packaging cavity diameter may underestimate the radiation level (overestimate the radial leakage).

The cross-section library used by the code should be applicable for shielding calculations. The code should account for subcritical multiplication and secondary gamma production unless these conditions have been otherwise appropriately considered (e.g., in the source-term specification).

5.4.2 Input and Output Data

This section should identify the key input data for the shielding calculations and show that information from the shielding models is properly input into the code. Depending on the type of computer code (e.g., point kernel, deterministic, Monte Carlo), key input data should include source terms, materials, package dimensions, convergence criteria, mesh size, neutrons per generation, and number of generations, with other input data included as appropriate. At least one representative input file and output file, or key sections of those files, should be included. This section should show that the code achieved proper convergence.

5.4.3 Flux-to-Dose-Rate Conversion

This section should include a tabulation of the flux-to-dose-rate conversion factors as a function of energy and should cite appropriate references to support the data. The flux-to-dose-rate conversion factors in ANSI/American Nuclear Society (ANS) 6.1.1-1977, "American National Standard for Neutron and Gamma-Ray Flux to Dose Factors" (Ref. 18), should be used for calculating the dose rates.

5.4.4 External Radiation Levels

This section should describe the results of the radiation analysis in detail. These results should agree with the summary tables in Section 5.1.2 and meet the limits in 10 CFR 71.47(a) or 10 CFR 71.47(b), as appropriate, and 10 CFR 71.51(a)(2) or Paragraphs 526, 530, 531, and 532 of TS-R-1 as referenced in Subsection 16(4) of the PTNS Regulations, Paragraph 572 of TS-R-1 as referenced in Subsection 15(5) of the PTNS Regulations, and Paragraphs 645, 646(b), and 656(b)(ii)(i) of TS-R-1 which are incorporated in Subsection 1(1) of the PTNS Regulations by reference to Paragraph 650 of TS-R-1. The locations of maximum dose rates for the analysis should be identified and sufficient data provided to show that the radiation levels are reasonable and their variations with location are consistent with the geometry and shielding characteristics of the package. The results should address normal and accident conditions.

The analysis should show that the locations selected are those of maximum dose rates. To determine maximum dose rates, radiation levels may be averaged over the cross-sectional area of a probe of reasonable size (See HPPOS-13, "Averaging of Radiation Levels over the Detector Probe Area," in NUREG/CR-5569, Revision 1, "Health Physics Positions Data Base," issued in 1992 (Ref. 19)). For packages with streaming paths or voids, averaging should not be used to reduce the radiation levels resulting from such features.

The external radiation levels should be reasonable and their variations with location should be consistent with the geometry and shielding characteristics of the package. For the purpose of 10 CFR 71.47(b) or Paragraph 572 of TS-R-1, the external surface is considered to be that part of the package that is shown in the drawings and has been demonstrated to remain in place under the normal conditions of transport. Personnel barriers and similar devices that are attached to the conveyance rather than the package may, however, qualify the vehicle as a closed vehicle.

The evaluation should address damage to the shielding under normal conditions of transport and hypothetical accident conditions. Applicants should verify that any damage under normal conditions of transport does not result in a significant increase in the external dose rates, as required by 10 CFR 71.43(f) and 10 CFR 71.51(a)(1) or Paragraph 646(b) of TS-R-1. Any increase should be explained and justified as not significant.

5.5 Appendix

The appendix should include a list of references, applicable pages from referenced documents, supporting information and analysis, test results, and other appropriate supplemental information.

6. Criticality Evaluation

This section of the application should identify, describe, discuss, and analyze the principal criticality safety design of the package, components, and systems important to safety and describe how the package complies with the requirements of 10 CFR 71.15, "Exemption from Classification as Fissile Material," 10 CFR 71.55, 10 CFR 71.59, and Paragraph 528 of TS-R-1 as referenced in Subsection 16(4) of the PTNS Regulations, Paragraph 671 of TS-R-1 which is incorporated in Paragraph 7(1)(a) of the PTNS Regulations by reference to Paragraph 813 of TS-R-1, Paragraph 672 of TS-R-1 as referenced in Subsection 1(1) of the PTNS Regulations and Paragraphs 673-682 of TS-R-1 which are incorporated in Subsection 1(1) of the PTNS Regulations by reference to Paragraph 672 of TS-R-1.

The following are exceptions from the requirements for fissile material packages.

The requirements for fissile material exemptions in 10 CFR 71.15 and Paragraph 672 of TS-R-1 are different. The applicant should comply with the most restrictive of the requirements regarding the fissile material exemption. For packages with fissile material meeting requirements in 10 CFR 71.15 and Paragraph 672 of TS-R-1, the packaging is exempted from the requirements in 10 CFR 71.55 and 10 CFR 71.59 and Paragraphs 528, 671, and 673–682 of TS-R-1. The fissile material should meet the provisions in both 10 CFR 71.15 and Paragraph 672 of TS-R-1 as follows:

1.   One of the requirements in 10 CFR 71.15(a) through (f); and

2.   One of the provisions in Paragraphs 672(a) through (d) in TS-R-1.

This section should address the structural and thermal effects on the packaging and its content under normal and hypothetical accident conditions in terms of material and geometry changes and subsequent effect on the criticality safety design. Any operational, fabrication, and maintenance requirements with respect to criticality safety importance for the package should be included in the application in Section 7, "Package Operations," and Section 8, "Acceptance Tests and Maintenance Program."

6.1 Description of Criticality Design

This section should describe the criticality design, which would include provisions in 10 CFR 71.31, "Contents of Application," 10 CFR 71.33, and Paragraphs 807 and 813 of TS-R-1 as referenced in Paragraph 7(1)(a) of the PTNS Regulations.

6.1.1 Design Features

This section should describe the design features of the package that are important for criticality control, including the following:

1.   Dimensions and tolerances of the containment system for fissile material;

2.   Structural components that maintain the fissile material and neutron poisons in a fixed position within the package and in a fixed position relative to each other;

59

3.   Location, dimensions, and concentration of neutron absorbing materials and moderating materials, including neutron poisons and shielding material;

4.   Dimensions and tolerances of floodable voids and flux traps within the package;

5.   Dimensions and tolerances of the overall package that affect the physical separation of the fissile material contents in package arrays; and

6.   Information on control rod assemblies, shrouds, or other fuel assembly components included with fresh fuel or SNF as applicable to the criticality evaluation.

All information presented in the text, drawings, figures, and tables should be consistent with each other and with that used in the criticality evaluation. The drawings are the authoritative source of dimensions, tolerances, and material composition of components important to criticality safety.

6.1.2 Summary Table of Criticality Evaluation

This section should provide a summary table of criticality analysis results for the package for the following cases, as described in Sections 6.4 through 6.6 of this document:

1.   A single package under the conditions of 10 CFR 71.55(b), (d), and (e) or Paragraphs 677, 678, and 679 of TS-R-1 which are incorporated in Subsection 1(1) of the PTNS Regulations by reference to Paragraph 672 of TS-R-1;

2.   An array of undamaged packages under the conditions of 10 CFR 71.59(a)(1) or Paragraph 681 of TS-R-1 which is incorporated in Subsection 1(1) of the PTNS Regulations by reference to Paragraph 672 of TS-R-1; and

3.   An array of damaged packages under the conditions of 10 CFR 71.59(a)(2) or Paragraph 682 of TS-R-1 which is incorporated in Subsection 1(1) of the PTNS Regulations by reference to Paragraph 672 of TS-R-1.

The maximum value of the effective neutron multiplication factor (k_{eff}), any stochastic uncertainty, the biases and associated uncertainties, and the number of packages evaluated in the arrays should be specified in the table. The table should also show that the sum of k_{eff}, two standard deviations, and the biases with their associated uncertainties adjustment does not exceed 0.95 for each case. Therefore, a package is considered subcritical, under the regulatory conditions, if it can satisfy the following relationship:

$$k_{eff} + 2\sigma + \Delta k_u \le 1 - \Delta k_m \qquad \text{Eq. 6-1}$$

Where

k_{eff} =   the calculated value obtained for the package or array of the packages

σ =   is the standard deviation of the K_{eff} value obtained with Monte Carlo analysis (the value for this parameter is set to zero if a deterministic method is used)

Δk_u =   an allowance for the calculation bias and uncertainty as discussed in Section 6.8 of this document

Δk_m = a required margin of subcriticality (minimum of 0.05 depending on the sensitivity of k_{eff} to the system parameter uncertainties)

Therefore, Equation 6-1 can be rewritten as follows:

$$K_{eff} + 2\sigma + \Delta k_u \le 0.95 \qquad \text{Eq. 6-2}$$

6.1.3 Criticality Safety Index

This section should provide the Criticality Safety Index (CSI) in accordance with 10 CFR 71.59(b) or Paragraph 528 of TS-R-1 as referenced in Subsection 16(4) of the PTNS Regulations, based on the number of packages evaluated in the arrays and show how it was calculated.

The CSI should be consistent with that reported in Section 1, "General Information," of the application. The value of N, which is the number of packages allowed for transport, should be specified.

6.2 Fissile Material Contents

This section should describe in detail the fissile material allowed in the package in accordance with 10 CFR 71.33 or Paragraphs 807 and 813 of TS-R-1 as referenced in Paragraph 7(1)(a) of the PTNS Regulations.

Specifications for the contents used in the criticality evaluation (e.g., fissile content, poison materials, content geometry) should be consistent with that described in Section 1 and throughout the application. Specifications relevant to the criticality evaluation should include fissile material mass, dimensions, enrichment, physical and chemical composition, density, moisture, and other characteristics dependent on the specific contents. Any differences from the specifications in Section 1 of the application or other sections should be clearly identified and justified. Because a partially filled container may allow more room for moderators (e.g., water), the most reactive case may be for a mass of fissile material that is less than the maximum allowable contents. Therefore, a minimum mass may have to be specified.

If the package is designed for multiple types of contents, the application may include a separate criticality evaluation and propose different criticality controls for each content type. Any assumptions that certain contents need not be evaluated because they are less reactive than evaluated contents should also be properly justified.

For SNF, specifications relevant to criticality evaluation should include the following:

1.   Types of fuel assemblies, plates, or rods (e.g., boiling-water reactor/pressurized-water reactor) and vendor/model as appropriate;

2.   Dimensions of fuel (including any annular pellets), cladding, fuel-cladding gap, pitch, and rod length;

3.   Number of rods or plates per assembly and locations of guide tubes and burnable poisons;

4.   Materials and densities;

5. Active fuel length;

6. Enrichment (variation by rod if applicable) before irradiation;

7. Chemical and physical form;

8. Mass of initial heavy metal per assembly or rod;

9. Number of fuel assemblies or individual rods per package; and

10. Other components when included in the criticality analysis or which have nonnegligible effect on k_{eff}.

The conditions of the SNF assemblies, including missing or replaced fuel rods, should be described. In general, the description of the contents should be sufficient to permit a detailed criticality evaluation of each type or to support a conclusion that certain types are bounded by the evaluations performed. If the contents include damaged fuel, the maximum extent of damage should be specified. Any cans or canisters used as part of the content for the packaging should be described.

6.3 General Considerations

This section should address general considerations used to evaluate the criticality of the package. These may apply to the criticality evaluations of a single package and arrays of packages under both normal conditions of transport and hypothetical accident conditions.

6.3.1 Model Configuration

This section should describe and provide sketches of the calculation model used in the calculations. The sketches should identify the materials used in all regions of the model. Any differences between the model and the actual package configuration should be identified and justification given that the model is conservative. In addition, the differences between the models for normal and accident conditions of transport should be clearly identified and justified.

Within the specified tolerance range, dimensions should be selected to result in the highest reactivity. For example, cavity sizes and poison thickness should be considered in the manner that maximizes reactivity.

Deviations from nominal design configuration should be considered. For example, the contents of a powder packaging can be positioned at varying locations and densities, the fuel assemblies might not always be centered in each basket compartment, and the basket might not be exactly centered in a spent fuel package. The relative location and physical properties of the contents within the packaging should be justified as those resulting in the maximum multiplication factor.

For fuel assembly packages, the fully flooded scenario should address preferential flooding and include flooding of the fuel-cladding gap. In addition, variable water density should be considered for possible system reactivity peaks.

Due to the capabilities of modern computer codes, homogeneous modeling should not be used. If homogenization is used in the model, this section should demonstrate that it is applied correctly and/or conservatively.

6.3.2 Material Properties

This section should provide the appropriate mass densities and atomic number densities for materials used in the models of the packaging and contents. Material properties should be consistent with the condition of the package under the tests specified in 10 CFR 71.71 and 10 CFR 71.73 or Paragraphs 719–724 and 726–729 of TS-R-1 which are incorporated in Subsection 1(4) of the PTNS Regulations by reference to Paragraph 716 of TS-R-1. Materials relied on for criticality control must remain in place and be effective.

No more than 75 percent of the specified minimum neutron poison concentration should generally be considered in the criticality evaluation unless a higher percentage can be justified.

The differences in material condition between normal conditions of transport and hypothetical accident conditions should be clearly identified. Materials relevant to the criticality design, such as poisons, foams, plastics, and other hydrocarbons, should specifically be addressed.

Neutron absorbers and moderators (e.g., poisons and neutron shielding) should be properly controlled during fabrication to meet their specified properties. Such information should be discussed in more detail in Section 8, "Acceptance Tests and Maintenance Program," of the application.

Materials should not degrade during the service life of the packaging to the point of affecting adversely the package performance. Specific controls should be in place to ensure effectiveness of the packaging during its service life. Such information should also be discussed in more detail in Section 8 of the application.

6.3.3 Computer Codes and Cross-Section Libraries

This section should describe the basic methods used to calculate the effective neutron multiplication constant of the package to demonstrate compliance with the fissile material package standards. This should address the following:

1. A description of the computer program and neutron cross-sections used;

2. The bases for selecting the specific program and cross-sections; and

3. Key input data for the criticality calculations, such as neutrons per generation, number of generations, convergence criteria, and mesh selection.

At least one representative input file and one output file (or key sections of these files) for a single package, undamaged array and damaged array should be included in the application. The calculation should properly converge and the calculated multiplication factors from the output files should agree with those reported in the evaluation.

6.3.4 Demonstration of Maximum Reactivity

This section should include a demonstration that the most reactive configuration of each case listed in Sections 6.4 through 6.6 of this document (single package, arrays of undamaged packages, and arrays of damaged packages) has been evaluated. All assumptions and approximations should be clearly identified and justified.

For packages with multiple SNF assembly types for the content, all assembly types should be analyzed or the bounding fuel assembly type should be justified and analyzed.

This section should identify the optimum combination of internal moderation (within the package) and interspersed moderation (between packages) as applicable. The following should be considered:

1. Moderation by water and any hydrogen-containing packaging materials, such as polyethylene;

2. Preferential flooding of different regions within the package; and

3. Partial loadings (i.e., fissile masses less than the maximum allowable mass).

6.3.5 Burnup Credit for Irradiated Fuel Packages

In designing the criticality control system for irradiated fuel packages, if the applicant is relying on the reduced reactivity of fuel assemblies caused by the depletion of fissile material and production of neutron-absorbing isotopes, biases and uncertainties in predicting isotopic inventory and reactivity of irradiated fuel assemblies in packages should be established. This is further explained in Section 6.8 of this document. The amount of irradiation necessary with respect to reactivity to load the fuels into packages should be presented as a function of initial enrichment and any restrictions on the conditions during irradiation. In addition, a final independent burnup verification measurement should be performed before loading irradiated fuel assemblies into packages for shipment.

Both 10 CFR 71.83, "Assumptions as to Unknown Properties," and Paragraph 673 in TS-R-1 which is incorporated in Subsection 1(1) of the PTNS Regulations by reference to Paragraph 672 of TS-R-1 require that when the properties of the fissile material are not known, those properties that give the maximum neutron multiplication will be assumed.

In addition, Paragraph 674 of TS-R-1 which is incorporated in Subsection 1(1) of the PTNS Regulations by reference to Paragraph 672 of TS-R-1 requires that:

1. The irradiation history that results in maximum neutron multiplication shall be assumed; and

2. Prior to shipment, a measurement for confirming the conservatism of the isotopic composition shall be made.

The applicant should include descriptions of benchmarking experiments used to establish biases and uncertainties associated with depletion and criticality models used for the irradiated fuel in the packaging, the bounding irradiation history parameter values, and the method for burnup verification measurement.

6.4 Single Package Evaluation

6.4.1 Configuration

This section should demonstrate that, as a design basis in accordance with 10 CFR 71.55(b) or Paragraph 677 of TS-R-1 which is incorporated in Subsection 1(1) of the PTNS Regulations by reference to Paragraph 672 of TS-R-1, a single package is subcritical under normal or accident conditions, whichever is more reactive, with the following assumptions:

1. Fissile material in its most reactive credible configuration consistent with the condition of the package and the chemical and physical form of the contents;

2. Water moderation to the most reactive credible extent, including water in-leakage to the containment system as specified in 10 CFR 71.55(b) or Paragraph 677 of TS-R-1; and

3. Full water reflection on all sides of the containment system as specified in 10 CFR 71.55(b)(3) or Paragraph 678 of TS-R-1 which is incorporated in Subsection 1(1) of the PTNS Regulations by reference to Paragraph 672 of TS-R-1 or reflection by the package materials, whichever results in the maximum reactivity.

---

10 CFR 71.55(c) provides an exception to 10 CFR 71.55(b) if a package incorporates "special design features" that ensure no single packaging error would permit water leakage.

Paragraph 677 of TS-R-1 permits routine approvals of package designs without assuming water in-leakage, provided the design incorporates special features to prevent such leakage. Paragraph 677(a) further defines "special features" as including "multiple high standard water barriers."

Because of this difference in regulations, this guide does not cover designs invoking 10 CFR 71.55(c) or Paragraph 677(a) of TS-R-1. Therefore, for a single package, water in-leakage has to be assumed at all times in and around the fissile material and the containment system to the most reactive extent.

Paragraph 678 of TS-R-1 requires the "confinement system" to be closely reflected by at least 20 cm (8 in.) of water or such greater reflection as may additionally be provided by the surrounding material of the packaging. The "confinement system" consists of those components of the packaging that maintain the geometric configuration of the fissile material inside the packaging with respect to criticality safety. In addition, 20 cm (8 in.) of water is believed to be equivalent to a "full reflection" as inferred from 10 CFR 71.55(b)(3). However, if a layer of water thicker than 20 cm (8 in.) makes the system more reactive, the larger water thickness should be used as part of the design basis.

In addition, this section should also demonstrate, in accordance with 10 CFR 71.55(d) or Paragraph 679(b) of TS-R-1 which is incorporated in Subsection 1(1) of the PTNS Regulations by reference to Paragraph 672 of TS-R-1, that the package content would be subcritical when the package is subjected to the normal conditions of transport.

Furthermore, this section should also demonstrate, in accordance with 10 CFR 71.55(e) or Paragraph 679(c) of TS-R-1 which is incorporated in Subsection 1(1) of the PTNS Regulations by reference to Paragraph 672 of TS-R-1, that the package content would be subcritical when the package is subjected to the accident conditions of transport.

6.4.2 Results

This section should present the results of the single package evaluation and should also address the additional specifications of 10 CFR 71.55(d)(2)–(d)(4) or Paragraph 679 of TS-R-1 which is incorporated in Subsection 1(1) of the PTNS Regulations by reference to Paragraph 672 of TS-R-1 under normal conditions of transport.

The results of the most reactive case for the single package analysis should be consistent with the information presented in the summary table as discussed in Section 6.1.2 of this document. If the package can be shown to be subcritical by reference to a standard, the standard should be applicable to the package conditions.

6.5 Evaluation of Package Arrays under Normal Conditions of Transport

6.5.1 Configuration

This section should evaluate, in accordance with 10 CFR 71.59(a)(1) or Paragraph 681 of TS-R-1 which is incorporated in Subsection 1(1) of the PTNS Regulations by reference to Paragraph 672 of TS-R-1, an array of 5N packages under normal conditions of transport. The evaluation should consider the following factors:

1.   The most reactive configuration of the array (e.g., pitch and package orientation) with nothing between the packages;

2.   The most reactive credible configuration of the packaging and its contents under normal conditions of transport (noting that if the water spray test has demonstrated that water would not leak into the package, water in-leakage need not be assumed for this case); and

3.   Full water reflection on all sides of a finite array.

6.5.2 Results

This section should present the results of the analyses for arrays and identify the most reactive array conditions. The results of the analysis should be consistent with the information presented in the summary table, as discussed in Section 6.1.2 of this document.

The appropriate N value should be used in determining the CSI. The appropriate N should be the smaller value that ensures subcriticality for 5N packages under normal conditions of transport or 2N packages under hypothetical accident conditions, as discussed in the next section.

6.6 Package Arrays under Hypothetical Accident Conditions

6.6.1 Configuration

This section should evaluate, in accordance with 10 CFR 71.59(a)(2) or Paragraph 682 of TS-R-1 which is incorporated in Subsection 1(1) of the PTNS Regulations by reference to Paragraph 672 of TS-R-1, an array of 2N packages under hypothetical accident conditions. The evaluation should consider the following factors:

1.   The most reactive configuration of the array (e.g., pitch, package orientation, and internal moderation);

2.   Optimum interspersed hydrogenous moderation;

3. The most reactive credible configuration of the packaging and its contents under hypothetical accident conditions, including in-leakage of water; and

4. Full water reflection on all sides of a finite array.

6.6.2 Results

This section should present the results of the analyses for arrays and identify the most reactive array conditions. The results of the analysis should be consistent with the information presented in the summary table, as discussed in Section 6.1.2 of this document.

The appropriate N value should be used in determining the CSI. The appropriate N should be the smaller value that ensures subcriticality for 5N packages under normal conditions of transport or 2N packages under hypothetical accident conditions.

6.7 Fissile Material Packages for Air Transport

6.7.1 Configuration

This section should evaluate a single package under the expanded accident conditions specified in 10 CFR 71.55(f) or Paragraph 680 of TS-R-1 which is incorporated in Subsection 1(1) of the PTNS Regulations by reference to Paragraph 672 of TS-R-1. The evaluation should consider the following factors:

1. The most reactive configuration of the contents and packaging under the expanded accident conditions;

2. Full water reflection; and

3. No water in-leakage.

6.7.2 Results

This section should present the results of the analyses for the single package and identify the most reactive contents and packaging conditions. The results of the analysis should be consistent with the information presented in the summary table, as discussed in Section 6.1.2 of this document.

6.8 Benchmark Evaluations

This section should describe the methods used to benchmark the criticality calculations. The computer codes for criticality calculations should be benchmarked against critical experiments. The same computer code, hardware, modeling methodology, and cross-section library used to calculate the effective multiplication factor values for the package should be used in the benchmark experiments. This section should present the results of calculations for selected critical benchmark experiments to justify the validity of the calculational method and neutron cross-section values used in the analysis.

The Organization for Economic Cooperation and Development's "International Handbook of Evaluated Criticality Safety Benchmark Experiments," issued in September 2005 (Ref. 20), provides a source for selecting applicable critical experiments in benchmarking the computer codes and cross-sections used for designing the packages.

6.8.1 Applicability of Benchmark Experiments

This section should describe selected critical benchmark experiments that were analyzed using the method and cross-sections given in Section 6.3 of this document. This section should show the applicability of the benchmarks in relation to the package and its contents, noting all similarities and resolving all differences. The benchmark experiments should have, to the maximum extent possible, the same material, neutron spectrum, and configuration as the package evaluations. Key package parameters that should be compared with those of the benchmark experiments include type of fissile material, enrichment, hydrogen-to-fissile atomic ratio (dependent largely on rod pitch and diameter for fuel assemblies), poisoning, reflector material, and configuration. References that give full documentation on these experiments should be provided. The applicant may use computer codes, such as "Tools for Sensitivity and Uncertainty Methodology Implementation" (TSUNAMI), developed by Oak Ridge National Laboratory as part of the SCALE 5.1 package (Ref. 21), to assess similarities of packages with critical systems for benchmarking purposes.

The overall quality of the benchmark experiments and any uncertainties in experimental data should be addressed. The uncertainties should be treated in a conservative manner. Results of the benchmark calculations as well as the actual nuclear and geometric input parameters used for those calculations should be provided.

6.8.2 Bias Determination

This section should present the results of the benchmark calculations and the method used to account for biases and uncertainties in calculations (i.e., Δk_u in Eq. 6-2) and the contribution from uncertainties in the experimental data. This section should show a sufficient number of appropriate benchmark experiments and demonstrate that the results of the benchmark calculations were appropriate to determine the bias for the package calculations. In search of biases, parameters such as pitch-to-rod diameter, assembly separation, and neutron absorber material should be considered. Statistical and convergence uncertainties should be addressed. Only negative biases (results that underpredict k_{eff}) should be considered, with positive bias results treated as zero bias.

In quantifying Δk_u for computer codes and cross-sections used in designing burnup-credit spent fuel packages, biases and uncertainties from both depletion and criticality computer codes should be included. In addition, biases associated with axial and horizontal variation of the burnup within a spent fuel assembly should be considered. Furthermore, the effects of reactor operating history on the reactivity of discharged spent fuel assemblies should be addressed.

6.9 Appendix

The appendix should include a list of references, applicable pages from referenced documents, justification of assumption or analytical procedures, test results, photographs, computer code descriptions, input and output files, and other supplemental information. Input files for representative or "most limiting" cases for a single package and arrays of damaged and undamaged packages should specifically be included.

7. **Package Operations**

This section of the application should describe the operations, as required by 10 CFR Part 71, (10 CFR 71.31(c), 10 CFR 71.35(c), 10 CFR 71.43(g), 10 CFR 71.47(b)-(d), 10 CFR 71.87, "Routine Determinations," and 10 CFR 71.89, "Opening Instructions") and Paragraph 807(d) of TS-R-1 as referenced in Paragraph 7(1)(*a*) of the PTNS Regulations, used to load a package and prepare it for transport, presenting the steps sequentially in the actual order in which they are performed. The operations should describe the fundamental steps needed to ensure that the package is properly prepared for transport, consistent with the package evaluation in Sections 2–6 of the application. The regulatory requirements for package operations are 10 CFR 71.87 and Paragraph 502 of TS-R-1 as referenced in Subsection 16(4) of the PTNS Regulations.

The package should be operated in accordance with detailed written procedures that are based on and consistent with the operations described in this section of the application. The package operations should be consistent with maintaining occupational radiation exposures as low as reasonably achievable (ALARA) as required by 10 CFR Part 20, "Standards for Protection Against Radiation," (Ref. 22) particularly 10 CFR 20.1101(b), or Paragraph 302 of TS-R-1.

The package operations presented in the application should not be detailed procedures that can be implemented without expansion. Rather, the package operations should be an outline that focuses upon those steps that are important to ensuring that the package is operated in a manner consistent with its evaluation for approval. Compliance with the submitted package operations will be included in the certificate by reference as a condition of approval. The application should only include those procedural details that are important to safety. Operational steps should normally be presented in sequential order, as applicable. Guidance on preparing both the detailed procedures and the package operations that are to be included with an application can be found in NUREG/CR-4775, "Guide for Preparing Operating Procedures for Shipping Packages" (Ref. 23) issued December 1988.

7.1 Package Loading

This section should describe loading-related preparations, tests, and inspections of the package, including the inspections made before loading the package to determine that the package is not damaged and that radiation and surface contamination levels are within allowable limits of the regulations.

7.1.1 Preparation for Loading

At a minimum, the operations for preparing the package for loading should ensure the following:

1. The package is loaded and closed in accordance with written instructions.

2. The contents are authorized by the certificate of compliance, including the use of a secondary container or containment as appropriate.

3. The use of the package complies with the conditions of approval in the certificate of compliance, including verification that the package meets the design identified in the approval, and that the required maintenance has been performed.

4. The package is in unimpaired physical condition.

5. For a fissile material shipment, any special controls and precautions for transport, loading, unloading, and handling and any appropriate actions in case of an accident or delay are provided to the carrier or consignee.

6. Any proposed special controls and precautions for handling the package are provided.

7. Any required moderator or neutron absorber is present and in proper condition in full conformance with the approved package design.

8. The package is properly labeled.

In addition, the operations should describe the inspection of seals, criteria for replacement, and repair processes, if applicable, as well as the inspection of each closure device and criteria for replacement.

7.1.2 Loading of Contents

At a minimum, the operations for loading the contents should describe how the contents are loaded and how the package is closed. The loading operations should ensure the following:

1. Any special handling equipment needed for loading and unloading is provided.

2. Any proposed special controls and precautions for loading and handling the package are provided.

3. Any moderator or neutron absorber, if specified, is present and in proper condition.

4. The package has been loaded and closed appropriately in accordance with the specified bolt torques and bolt-tightening sequences.

5. If appropriate, methods to drain and dry the package are described, the effectiveness of the proposed methods is discussed, and vacuum drying criteria are specified.

6. For spent fuel packages, special controls and precautions regarding damaged fuel are provided.

7. Each closure device of the package, including any specified seals, is properly installed and secured and free of defects.

For SNF packages, if the applicant has designed the criticality control system of the packages based on the reduced reactivity of irradiated nuclear fuel assemblies due to burnup, a description of the steps involved in the measurement to independently verify the burnup of the irradiated nuclear fuel should be provided in this section. The description should include the measurement technique, type of device, the measured parameter(s), number of measurements, and the criterion for an acceptable error band on the measurement device.

7.1.3 Preparation for Transport

The operations for preparing the package for transport should address radiation and contamination surveys of the package, leakage testing of the package, measurement of the package surface temperature, package tie-down, and the application of tamper-indicating devices and appropriate marking and labeling. At a minimum, the preparation for transport operations should ensure the following:

1.   The level of non-fixed (removable) radioactive contamination on the external surfaces of each package offered for shipment is ALARA, and within the limits specified in DOT regulation 49 CFR 173.443, "Contamination Control" (Ref. 24) or Paragraph 508 of TS-R-1 as referenced in Subsection 16(4) of the PTNS Regulations.

2.   Radiation survey requirements of the package exterior are described to ensure that limits specified in 10 CFR 71.47 or Paragraph 531 of TS-R-1 as referenced in Subsection 16(4) of the PTNS Regulations, and Paragraph 572 of TS-R-1 as referenced in Subsection 15(5) of the PTNS Regulations are met.

3.   The temperature of the package exterior is within the limits specified in 10 CFR 71.43(g) or Paragraphs 617, 652, and 662 of TS-R-1 which are incorporated in Subsection 1(1) of the PTNS Regulations by reference to Paragraph 650 of TS-R-1.

4.   For Type B packages for non-special (normal) form radioactive material, pre-shipment leakage testing is performed as specified in ANSI N14.5 (Ref. 15) or ISO 12807 (Ref. 16). This test is non-quantitative, and is used to demonstrate no leakage at a specified test sensitivity.

5.   A tamper-indicating device is incorporated which, while intact, indicates that the package has not been opened by unauthorized persons.

6.   Any system for containing liquid is adequately sealed and has adequate space or other specified provision for expansion of the liquid.

7.   A check is made to ensure that any pressure relief device is operable and properly set.

8.   Any structural part of the package that could be used to lift or to tie down the package during transport is rendered inoperable for that purpose unless it satisfies the design requirements for lifting or tie-down attachments as required by 10 CFR 71.45, "Lifting and Tie-Down Standards for All Packages," or Paragraph 608 of TS-R-1 which is incorporated in Subsection 1(1) of the PTNS Regulations by reference to Paragraph 650 of TS-R-1.

9.   Any proposed special controls and precautions for transport and handling and any proposed special controls in case of accident or delay are specified.

10.   Marking and labeling of the package are done in accordance with 49 CFR 172.310, "Class 7 (Radioactive) Materials," and 49 CFR 172.403, "Class 7 (Radioactive) Material," (Ref. 24) or Paragraphs 535-540 and 542-546 of TS-R-1 as referenced in Subsection 16(4) of the PTNS Regulations.

11.   Written instructions to the carrier are provided for packages that require exclusive use shipment because of external radiation levels.

12.   Before delivery of a package to a carrier for transport, the licensee has sent or made available to the consignee any special instructions needed to safely open the package.

7.2 Package Unloading

This section should include inspections, tests, and special preparations of the package for unloading. As applicable, this section should also describe the operations used to ensure safe removal of fission gases, contaminated coolant, and solid contaminants.

7.2.1 Receipt of Package from Carrier

The process for receiving the package should address radiation and contamination surveys and inspection of the tamper-indicating device. This section should also describe any proposed special controls and precautions for handling and unloading.

> The U.S. regulations require the receiver to conduct a radiation and contamination survey of packages in accordance with 10 CFR 20.1906, "Procedures for Receiving and Opening Packages." Subsection 21(3) of the Canadian PTNS regulations requires that the consignee verify whether the package is damaged, shows evidence of having been tampered with, has any portion of the fissile material outside the confinement system, and has any portion of the contents outside the containment system.

The operations should be presented sequentially in the order of performance for receiving the package from the carrier. At a minimum, the receipt operations should ensure the following:

1.   The requirements of 10 CFR 20.1906 or PTNS Subsection 21(3) are met.

2.   The package is examined for visible external damage.

3.   Steps to define actions to be taken when the tamper-indicating device is not intact or surface contamination or radiation survey levels are too high are provided.

4.   A list of any special handling equipment needed for unloading and handling the package is provided.

5.   Any proposed special controls and precautions for unloading and handling the package are provided.

6.   Procedures controlling the radiation level limits on unloading operations are provided.

7.   Procedures for the safe removal of fission gases, contaminated coolants, and solid contaminants, if any, are provided.

7.2.2 Removal of Contents

This section should describe the appropriate operations and method for opening and removing contents from the package. The operations should be presented sequentially in the order of performance for removing the contents following package receipt. At a minimum, the operations should ensure the following:

1.   The closure is removed appropriately.

2.   The contents are removed appropriately.

3. Verification is made that the contents are completely removed.

7.3 Preparation of Empty Package for Transport

This section should describe the inspections, tests, and special preparations needed to ensure that the packaging is verified to be empty and is properly closed and that the radiation and contamination levels are within allowable limits. In addition, this section should address the appropriate requirements of 49 CFR 173.428, "Empty Class 7 (Radioactive) Materials Packaging," (Ref. 24) or the requirements in Paragraph 520 of TS-R-1 as referenced in Subsection 16(4) of the PTNS Regulations. The operations should ensure the following:

1. The packaging is empty.

2. Appropriate inspections and tests of the package are performed before transport to ensure that the requirements of 10 CFR 71.87(i) or Paragraph 508 of TS-R-1 as referenced in Subsection 16(4) of the PTNS Regulations are met.

3. Special preparations of the packaging to ensure that the interior of the packaging is properly decontaminated and closed in accordance with the requirements of 49 CFR 173.428 (Ref. 24) or Paragraph 520 of TS-R-1 are described.

7.4 Other Operations

This section should include the provisions for any special operational controls (e.g., route, weather, shipping time restrictions).

7.5 Appendix

The appendix should include a list of references, applicable pages from referenced documents, detailed descriptions and analysis of processes or protocols, graphic presentations, test results, and other appropriate supplemental information.

8. Acceptance Tests and Maintenance Program

This section of the application should describe the acceptance tests and maintenance program, as required by Paragraph 807(d) of TS-R-1 as referenced in Paragraph 7(1)(a) of the PTNS Regulations, to be used for the packaging in compliance with Subpart G, "Operating Controls and Procedures," of 10 CFR Part 71, and Paragraph 501 of TS-R-1 as referenced in Subsection 16(4) of the PTNS Regulations. The acceptance tests and maintenance program will be included in the certificate by reference as a condition of approval.

8.1 Acceptance Tests

This section should describe the tests, as required in 10 CFR 71.85, "Preliminary Determinations," and Paragraph 501 of TS-R-1 as referenced in Subsection 16(4) of the PTNS Regulations, to be performed before the first use of each packaging. Each test and its acceptance criteria should be described. The acceptance tests should confirm that each packaging is fabricated in accordance with the drawings referenced in the package approval. The specificity of the information provided on the acceptance tests should be sufficient to verify the adequacy of the packaging.

Paragraph 501 of TS-R-1 and 10 CFR 71.85 contain slightly different requirements for acceptance testing. 10 CFR 71.85 requires that it be determined that there are no cracks, pinholes, uncontrolled voids, or other defects that could significantly reduce the effectiveness of the packaging. Paragraph 501 of TS-R-1 requires that the effectiveness of its shielding and containment and, where necessary, the heat transfer characteristics and the effectiveness of the confinement system, are within the limits for the approved design. If the design pressure of the containment system exceeds 35 kPa (5 lbf/in.^{2}) gauge, Paragraph 501 requires it be ensured that the containment system of each package conforms to the approved design requirements relating to the capability of that system to maintain its integrity under that pressure. 10 CFR 71.85 more specifically requires a test of the containment system at an internal pressure at least 50 percent higher than the maximum normal operating pressure.

Additional regulatory requirements of 10 CFR Part 71 applicable to the acceptance tests include the following:

1. The application must identify codes, standards, and the specific provisions of the QA program used for the acceptance testing of the packaging (10 CFR 71.31(c) and 10 CFR 71.37(b)).

2. Before first use, the fabrication of each packaging must be verified to be in accordance with the approved design (10 CFR 71.85(c)).

3. Before first use, each packaging must be conspicuously and durably marked with its model number, serial number, gross weight, and a package identification number assigned by the NRC (10 CFR 71.85(c)).

4. The licensee must perform any tests deemed appropriate by the NRC (10 CFR 71.93(b)).

5.   Before first use, if applicable, the amount and distribution of the neutron-absorbing materials and moderators must be verified to meet the design specification (10 CFR 71.87(g)).

NUREG/CR-3854 (Ref. 10) provides additional guidance on acceptance tests.

8.1.1 Visual Inspections and Measurements

This section should describe the visual inspections to be performed and the intended purpose of each inspection. The criteria for acceptance of each inspection, as well as the action to be taken if noncompliance is encountered, should be described. The inspections should verify that the packaging has been fabricated and assembled in accordance with the drawings and that all dimensions and tolerances specified on the drawings are confirmed by measurement.

8.1.2 Weld Examinations

This section should describe welding examinations used to verify fabrication in accordance with the drawings, codes, and standards specified in the application. The locations, types, and sizes of welds should be confirmed by measurement. Other applicable specifications for weld performance, nondestructive examination, and acceptance should be identified.

Additional guidance on welding criteria is provided in NUREG/CR-3019 (Ref. 11).

8.1.3 Structural and Pressure Tests

This section should identify and describe the structural or pressure tests. Such tests should comply with 10 CFR 71.85(b) and Paragraph 501(a) of TS-R-1 (see shaded box in Section 8.1 of this document), as well as applicable codes or standards specified. The sensitivity of the tests, and the actions taken when the prescribed criteria are not met, should be specified. Structural testing of lifting trunnions should be conducted in accordance with NUREG-0612, "Control of Heavy Loads at Nuclear Power Plants, Resolution of Generic Technical Activity A-36," issued July 1980 (Ref. 25), ANSI N14.6-1993, "Radioactive Materials—Special Lifting Devices for Shipping Containers Weighing 10,000 Pounds (4,500 kg) or More" (Ref. 26), or other appropriate specification.

8.1.4 Leakage Tests

This section should describe the leak tests to be performed on the containment vessel, as well as auxiliary equipment. The leakage tests should comply with ANSI N14.5 (Ref. 15) or ISO 12807 (Ref. 16). The acceptable leakage criterion should be consistent with that identified in Section 4, "Containment" of this document. The sensitivity of the tests should be specified, including the basis of this value, the criteria for acceptance, and the action to be taken if the criteria are not met. Methods for leak testing of all containment seals and penetrations, including drain and vent ports, should be described. The maximum allowable leakage rate and the minimum test sensitivity should be specified for each type of test (e.g., fabrication, maintenance, periodic, and preshipment leakage rate tests).

8.1.5 Component and Material Tests

This section should specify the appropriate tests and acceptance criteria for components that affect package performance. In addition, this section should specify test sensitivity, if applicable, provide acceptance criteria, and describe the action to be taken if those criteria are not met. Applicable QA procedures should be described to justify that the tested components are equivalent to the components that will be used in the packaging.

This section should also specify the appropriate tests and acceptance criteria for packaging materials. Tests should include those components, such as gaskets, under conditions that simulate the most severe service conditions under which they are to perform, including performance at pressure and under high and low temperatures. Tests for neutron absorbers (e.g., boron) and insulating materials (e.g., foams, fiberboard) should ensure that minimum specifications for density and isotopic content are achieved. In addition, tests that demonstrate the ability of the materials to meet the performance specifications shown on the engineering drawings should be described.

8.1.6 Shielding Tests

This section should specify the appropriate shielding tests for both neutron and gamma radiation. These tests and acceptance criteria should be sufficient to ensure that no defects, voids, or streaming paths exist in the shielding.

8.1.7 Thermal Tests

This section should specify the appropriate tests to demonstrate the heat transfer capability of the packaging. These tests should confirm that the heat transfer performance determined in the thermal evaluation (Section 3 of the application) is achieved in the fabrication process.

8.1.8 Miscellaneous Tests

This section should describe any additional tests to be performed before the packaging is used.

8.2 Maintenance Program

This section should describe the maintenance program used to ensure continued performance of the packaging as required by Paragraph 310(b) of TS-R-1 as referenced in Paragraph 13(*a*) of the PTNS Regulations, and Paragraph 807(d) of TS-R-1 as referenced in Paragraph 7(1)(*a*) of the PTNS Regulations. This program should include periodic testing, inspection, and replacement schedules, as well as criteria for replacement and repair of components and subsystems on an as-needed basis. The information provided on the maintenance program should be sufficient to demonstrate that the performance of the packaging will not degrade during its service life. The specificity of the information should be consistent with the importance of the maintenance in ensuring this continued performance.

Regulatory requirements of 10 CFR Part 71 applicable to the maintenance program include the following:

1.   The application must identify codes, standards, and the specific provisions of the QA program used for the maintenance program for the packaging (10 CFR 71.31(c) and 10 CFR 71.37(b)).

2.   The packaging must be maintained in unimpaired physical condition except for superficial defects such as marks or dents (10 CFR 71.87(b)).

3. The presence of any moderator or neutron absorber, if required, in a fissile material package must be verified before each shipment (10 CFR 71.87(g)).

4. The licensee must perform any tests deemed appropriate by the NRC (10 CFR 71.93(b)).

8.2.1 Structural and Pressure Tests

This section should identify and describe any periodic structural or pressure tests. Such tests should comply with 10 CFR 71.85(b), as well as applicable codes, standards, or other procedures specified in the application. Periodic structural testing of lifting trunnions should be conducted in accordance with NUREG-0612 (Ref. 25), ANSI N14.6 (Ref. 26), or other appropriate specification.

8.2.2 Leakage Tests

This section should describe the tests to be performed, the frequency with which those tests are performed, and the sensitivity of each test. For most systems, this description should include a test of the package before each shipment and annually. The leakage tests should be in compliance with ANSI N14.5 (Ref. 15) or ISO 12807 (Ref. 16). The acceptable leakage criterion should be consistent with that identified in Section 4, "Containment," of the application. In general, this section should specify that elastomeric seals should be replaced and leak tested within the 12-month period before shipment and that metallic seals should also be replaced and tested before each shipment.

8.2.3 Component and Material Tests

This section should describe the periodic tests and replacement schedules for components. Appropriate tests and their acceptance criteria to ensure packaging effectiveness for each shipment should be specified. Any process that could result in deterioration of packaging materials, including loss of neutron absorbers, reduction in hydrogen content of shields, and density changes of insulating materials should be addressed. Replacement intervals for components, such as bolts, that are susceptible to fatigue should be specified.

8.2.4 Thermal Tests

This section should describe the periodic tests used to ensure heat-transfer capability during the service life of the packaging. This section should describe periodic thermal tests, similar to the acceptance tests discussed in Section 8.1.7 of this document, and the interval for the tests, which is typically 5 years.

8.2.5 Miscellaneous Tests

Any additional tests to be performed periodically on the package or its components should be described.

8.3 Appendix

The appendix should include a list of references, applicable pages from referenced documents, test data, reports, and other appropriate supplemental information.

9. Quality Assurance

In this section, the applicant should describe the QA program, as required in 10 CFR 71.37 "Quality Assurance," or Paragraph 310 of TS-R-1 as referenced in Paragraph 13(*a*) of the PTNS Regulations. The applicant should demonstrate that effective and adequate QA programs are specified and established for those aspects of the design, manufacture, testing, documentation, use, maintenance, and inspection of packages, transport, and storage in transit for which the applicant is responsible.

IAEA Safety Series No. 113, "Quality Assurance for the Safe Transport of Radioactive Material" issued in 1994 (Ref. 27), provides guidance on the minimum requirements for an acceptable QA program. For package designers, the following program elements are required:

1.   QA programs;

2.   Organization;

3.   Document control;

4.   Design control;

5.   Procurement control;

6.   Nonconformance control;

7.   Corrective actions;

8.   Records;

9.   Staff and training; and

10.   Audits.

> Packages originally certified in Canada must meet the CNSC QA requirements described in Section 9.2 of this document, while packages originally certified in the United States must meet the NRC QA requirements outlined in Section 9.1.
>
> Validation requests should reference the applicable QA program approved with the original certification. Additional information may be requested to determine whether the referenced QA program is sufficient for package validation purposes.

9.1 U.S. Quality Assurance Program Requirements

For U.S. packages, NRC regulations require that, before the fabrication, testing, modification, or use of any package, the applicant obtain NRC approval of its QA program. The applicant must file a description of its QA program, including a discussion of which requirements of 10 CFR Part 71, Subpart H, "Quality Assurance," are applicable and how they will be satisfied. A previously approved QA program that satisfies the applicable criteria is acceptable. A QA program that NRC approves under Appendix B, "Quality Assurance Criteria for Nuclear Power Plants and Fuel Reprocessing Plants," to 10 CFR Part 50, "Domestic Licensing of Production and Utilization Facilities" (Ref. 28), or Subpart G, "Quality Assurance," of 10 CFR Part 72, "Licensing Requirements for the Independent Storage of Spent Nuclear Fuel, High-Level Radioactive Waste, and Reactor Related Greater than Class C Waste" (Ref. 29) is equivalent to a QA program that the staff approves under 10 CFR Part 71. The NRC has also endorsed the use of ANSI/ASME NQA-1-1983, "Quality Assurance Program Requirements for Nuclear Power Facilities" (hereafter referred to as NQA-1) (Ref. 30), as a standard that, when properly applied and supplemented (as necessary) to meet all applicable criteria, should result in the development of a QA program that is acceptable to the NRC staff.

Applicants should be aware that the QA regulations in 10 CFR Part 71 include requirements that other standards may not fully address. In general, programs based on NQA-1 (Ref. 30), Safety Series No. 113 (Ref. 27), or the ISO 9000, "Quality Management Issues," issued in 2000 (Ref. 31) standards will require supplementation in order to address all Subpart H regulations. The only exception is the 1983 revision of NQA-1, which the NRC has endorsed in its entirety. Without supplementation, the NRC may require the QA program user to submit additional information regarding how the applicable Subpart H regulations will be met.

Subpart H of 10 CFR Part 71 establishes QA requirements that apply to designing, purchasing, fabricating, handling, shipping, storing, cleaning, assembling, inspecting, testing, operating, maintaining, repairing, and modifying packaging components that are important to safety. To meet those requirements, licensees should control the quality of each of the above activities using a graded approach; that is, the QA effort that a licensee expends on an activity should be consistent with the importance to safety of the associated structures, systems, and components. Structures, systems, and components important to safety refer to the features of a Type B or fissile material package that are intended to perform the following functions:

1. Maintain the conditions required to safely transport the package contents;

2. Prevent damage to the package during transport; or

3. Provide reasonable assurance that the radioactive contents can be received, handled, transported, and retrieved without undue risk to the health and safety of the public or the environment.

NUREG/CR-6407, "Classification of Transportation Packaging and Dry Spent Fuel Storage System Components According to Importance to Safety," issued in 1996 (Ref. 32), provides a graded approach for classification of components used in transportation packaging according to importance to safety.

For U.S. packages, the QA program description should address the following elements from 10 CFR Part 71, Subpart H:

1. QA organization;

2. QA program;

3. Package design control;

4. Procurement document control;

5. Instructions, procedures, and drawings;

6.	Document control;

7.	Control of purchased material, equipment, and services;

8.	Identification and control of materials, parts, and components;

9.	Control of special processes;

10.	Internal inspection;

11.	Test control;

12.	Control of measuring and test equipment;

13.	Handling, storage, and shipping control;

14.	Inspection, test, and operating status;

15.	Nonconforming materials, parts, or components;

16.	Corrective action;

17.	QA records; and

18.	Audits.

9.2 Canadian Quality Assurance Program Requirements

For Canadian packages, the applicant should include a copy of the QA program that meets the requirements of Paragraph 310 of TS-R-1 as referenced in Paragraph 13(*a*) of the PTNS Regulations. IAEA Safety Series 113 (Ref. 27), or standards from the Canadian Standards Association, or the ISO may be followed in establishment of the program. A previously approved QA program that satisfies the applicable criteria is acceptable.

G. REFERENCES

1. Canadian Nuclear Safety Commission, "Packaging and Transport of Nuclear Substances (PTNS) Regulations," SOR/2000-208.

2. *U.S. Code of Federal Regulations*, "Packaging and Transportation of Radioactive Material," Part 71, Chapter I, Title 10, "Energy."

3. International Atomic Energy Agency, "Regulations for the Safe Transport of Radioactive Material," 1996 Edition (Revised), Safety Standards Series No. TS-R-1, Vienna, 2000

4. *U.S. Code of Federal Regulations*, "Rules of Practice for Domestic Licensing Proceedings and Issuance of Orders," Part 2, Chapter I, Title 10, "Energy."

5. United Nations, "Recommendations on the Transport of Dangerous Goods," Fourteenth Revised Edition, United Nations, New York and Geneva, 2005.

6. U.S. Nuclear Regulatory Commission, "Engineering Drawings for 10 CFR Part 71 Package Approvals," NUREG/CR-5502, May 1998.

7. U.S. Nuclear Regulatory Commission, "Design Criteria for the Structural Analysis of Shipping Cask Containment Vessels," Regulatory Guide 7.6, March 1978.

8. U.S. Nuclear Regulatory Commission, "Load Combinations for the Structural Analysis of Shipping Casks for Radioactive Material," Regulatory Guide 7.8, March 1989.

9. American Society of Mechanical Engineers, *Boiler and Pressure Vessel Code*, Section III, Division 3, "Containment Systems and Transport Packagings for Spent Nuclear Fuel and High Level Radioactive Waste," New York, 2007.

10. U.S. Nuclear Regulatory Commission, "Fabrication Criteria for Shipping Containers," NUREG/CR-3854, March 1985.

11. U.S. Nuclear Regulatory Commission, "Recommended Welding Criteria for Use in the Fabrication of Shipping Containers for Radioactive Materials," NUREG/CR-3019, March 1985.

12. U.S. Nuclear Regulatory Commission, "Fracture Toughness Criteria of Base Material for Ferritic Steel Shipping Cask Containment Vessels with a Maximum Wall Thickness of 4 Inches (0.1 m)," Regulatory Guide 7.11, June 1991.

13. U.S. Nuclear Regulatory Commission, "Fracture Toughness Criteria of Base Material for Ferritic Steel Shipping Cask Containment Vessels with a Maximum Wall Thickness Greater than 4 Inches (0.1 m) But Not Exceeding 12 Inches (0.3 m)," Regulatory Guide 7.12, June 1991.

14. American National Standards Institute, "Characterizing Damaged Spent Nuclear Fuel for the Purpose of Storage and Transport," ANSI N14.33-2005, New York, 2005.

15.   American National Standards Institute, "American National Standard for Radioactive Materials—Leakage Tests on Packages for Shipment," ANSI N14.5-1997, New York, 1997.

16.   International Standardization Organization, "Safe Transport of Radioactive Materials—Leakage Testing on Packages," ISO 12807:1996(E), Switzerland, 1996.

17.   U.S. Nuclear Regulatory Commission, "Containment Analysis for Type B Packages Used to Transport Various Contents," NUREG/CR-6487, November 1996.

18.   American National Standards Institute/American Nuclear Society, "American National Standard for Neutron and Gamma-Ray Flux to Dose Factors," ANSI/ANS 6.1.1, La Grange Park, Illinois, 1977.

19.   U.S. Nuclear Regulatory Commission, HPPOS-13, "Averaging of Radiation Levels over the Detector Probe Area," in NUREG/CR-5569, Rev. 1, "Health Physics Positions Data Base," 1992.

20.   Organization for Economic Cooperation and Development, "International Handbook of Evaluated Criticality Safety Benchmark Experiments," NEA/NSC/DOC(95)03, Vols. I–VIII, OECD, September 2005.

21.   Oak Ridge National Laboratory, "Standardized Computer Analyses for Licensing Evaluation (SCALE), Tools for Sensitivity and Uncertainty Methodology Implementation (TSUNAMI)," Rev. 5.1, December 2005.

22.   *U.S. Code of Federal Regulations*, "Standards for Protection Against Radiation," Part 20, Chapter I, Title 10, "Energy."

23.   U.S. Nuclear Regulatory Commission, "Guide for Preparing Operating Procedures for Shipping Packages," NUREG/CR-4775, December 1988.

24.   *U.S. Code of Federal Regulations*, "Hazardous Materials and Oil Transportation," Parts 100–185, Chapter 1, Title 49, "Transportation," October 2007.

25.   U.S. Nuclear Regulatory Commission, "Control of Heavy Loads at Nuclear Power Plants, Resolution of Generic Technical Activity A-36," NUREG-0612, July 1980.

26.   American National Standards Institute, "Radioactive Materials—Special Lifting Devices for Shipping Containers Weighing 10,000 Pounds (4500 kg) or More," ANSI N14.6-1993, New York, 1993.

27.   International Atomic Energy Agency, "Quality Assurance for the Safe Transport of Radioactive Material," Safety Series 113, IAEA, Vienna, 1994.

28.   *U.S. Code of Federal Regulations*, "Domestic Licensing of Production and Utilization Facilities," Part 50, Chapter I, Title 10, "Energy."

29.   *U.S. Code of Federal Regulations*, "Licensing Requirements for the Independent Storage of Spent Nuclear Fuel, High-Level Radioactive Waste, and Reactor-Related Greater Than Class C Waste," Part 72, Chapter I, Title 10, "Energy."

30. American National Standards Institute/American Society of Mechanical Engineers, "Quality Assurance Program Requirements for Nuclear Power Facilities," ANSI/ASME NQA-1-1983, New York, 1983.

31. International Standardization Organization, "Quality Management Systems," ISO 9000, Geneva, Switzerland, 2000.

32. U.S. Nuclear Regulatory Commission, "Classification of Transportation Packaging and Dry Spent Fuel Storage System Components According to Importance to Safety," NUREG/CR-6407, 1996.

H. GLOSSARY

buckling—In engineering, buckling is a failure mode characterized by a sudden failure of a structural member (e.g., a containment shell) subjected to high compressive stresses, where the actual compressive stresses at failure are smaller than the ultimate compressive stresses that the material is capable of withstanding. This mode of failure is also described as failure caused by elastic instability.

burnup—The induced nuclear transformation of atoms during reactor operation.

criticality safety index—A number assigned to a package, overpack, or freight container containing fissile material that provides control over the accumulation of packages, overpacks, or freight containers containing fissile material.

crud—A colloquial term for corrosion and wear products (e.g., rust particles) that become radioactive (i.e., activated) when exposed to radiation. Because the activated deposits were first discovered at Chalk River, a Canadian nuclear plant, "crud" has been used as shorthand for Chalk River unidentified deposits.

fissile material—Uranium-233, uranium-235, plutonium-239, plutonium-241, or any combination of these radionuclides, with the exception of the following:

- natural uranium or depleted uranium that is not irradiated
- natural uranium or depleted uranium that has been irradiated in thermal reactors only

Monte Carlo analysis—A method that uses computational algorithms for simulating the behavior of various physical and mathematical systems by the use of random numbers.

point-kernel technique—A technique based on an analytic point source solution where the unattenuated flux at any distance r from the source point is proportional to the source rate divided by $4\pi r^2$. Attenuations are treated in an approximate manner through the use of built-in attenuation coefficients and buildup factors.

Poisson's Ratio—The ratio of lateral strain to longitudinal strain, within the elastic range, for axially loaded specimens. Values of Poisson's ratio are required for structural analysis.

pyrophoricity—The property of a material or substance that ignites spontaneously when exposed to air or when rubbed or struck.

slapdown—The secondary impact resulting from the package initially impacting on a corner or an edge.

spent nuclear fuel—Nuclear reactor fuel that has been irradiated to the extent that it can no longer effectively sustain a chain reaction because its fissionable isotopes have been partially consumed and fission-product poisons have accumulated in it.